FOOTPRINTS

FOOTPRINTS

IN SEARCH OF FUTURE FOSSILS

DAVID FARRIER

4th ESTATE • London

4th Estate
An imprint of HarperCollins*Publishers*
1 London Bridge Street
London SE1 9GF

www.4thEstate.co.uk

First published in Great Britain in 2020 by 4th Estate
First published in the United States in 2020 by Farrar, Straus and Giroux

1

A catalogue record for this book is
available from the British Library

ISBN 978-0-00-828634-7 (hardback)
ISBN 978-0-00-831787-4 (trade paperback)

Designed by Gretchen Achilles

Set in Adobe Garamond
Printed and bound in Great Britain by
CPI Group (UK) Ltd, Croydon

MIX
Paper from
responsible sources
FSC™ C007454

This book is produced from independently certified FSC™ paper
to ensure responsible forest management.

Find out more about HarperCollins and the environment at
www.harpercollins.co.uk/green

ROYAL BOROUGH OF GREENWICH

Follow us on twitter 🐦 @greenwichlibs

Please return by the last date shown

5/21		

WITHDRAWN

For Isaac and Annie

CONTENTS

FOOTPRINTS

TRACES OF A HAUNTED FUTURE

England's eastern edge is slowly being reclaimed by the sea. At a rate of around two metres per year, the tides carry away the shallow cliffs that form the East Anglian coastline. Seasonal storms gouge away the earth, mostly glacial till that was deposited when ice sheets reached as far as southern Britain 450,000 years ago; like a cheaply made wall, the coast is vulnerable to erosion and sudden collapses. One night in 1845, a farmer ploughed a twelve-acre field near Happisburgh, in Norfolk, and retired to bed with the soil ready for planting the following morning. When he woke, the field had vanished. Coastal defences built after devastating floods in 1953, which killed over three hundred people, have long since crumbled. Buildings that once stood out of sight of the coast now huddle within its reach, and homeowners watch anxiously as the margin creeps closer, eating up their lovingly tended gardens inch by inch. Occasionally, a house tumbles into the sea. The ground beneath your feet feels provisional, as though you are standing on borrowed time.

Sometimes, though, the sea gives something back. In May 2013, a spring storm uncovered the oldest traces of human passage outside Africa on Happisburgh's claggy foreshore. Heaving seas had stripped away the sand behind the dilapidated postwar flood defences to reveal a section of laminated silt flecked with dozens of lozenge-shaped hollows. The scooped depressions were the 850,000-year-old fossil footprints left by a group of early humans, *Homo antecessor*, moving along the muddy banks of an ancient river. The differently sized footprints suggested that it was a group of mixed age, adults and children, headed in a southerly direction. At the time, it was an estuarine landscape, populated by pine, spruce, and birch forests interspersed with open areas of heath and grassland. Photographs of the footprints resemble a step map of a frantic dance floor. The busy play of feet suggests a neighbourly scene: adults pausing to coax tired children or turning to check the horizon for predators; arms raised to indicate points of interest or offer an encouraging hand on the shoulder. Some impressions were so well preserved that you could see the outlines of individual toes.

Briefly, startlingly, this small party of hominids walked out of the deep past and into the present. They disappeared almost as fast: within two weeks, the tide had washed away every print.

Ancient footprints, like burrows, tracks and tooth marks, are known as trace fossils. Unlike fossilized remains, they speak of life rather than death. Though bodiless, they bear witness to a departed body's weight, gait, and habits, telling stories about how ancient lives were lived. Trace fossils like the Happisburgh footprints are an accidental memory; where the group came from and where they were heading are beyond our knowing. But the prints offer an enchanting glimpse of ancestors whose past seems to brush against our present, whose step into our time seems like an invitation to join in a mysterious journey. Even in photographs they conjure the uncanny sensation that the group has only just left, their prints fresh and glistening – that we could catch them, if only we hurried.

In terms of traces left by early humans, the Happisburgh footprints are relatively young. The oldest known hominid prints were made 3.6 million years ago, in volcanic ash at Laetoli in what is now the Ngorongoro Conservation Area, in Tanzania. They were discovered in 1976 and embraced as marks of a Pliocene 'first family', making their way like Milton's Adam and Eve, 'hand in hand with wand'ring steps and slow'. When the deep past arrives in the present, it is often surprising. The Laetoli footprints were found when, in high spirits during a break in their labours, a team of palaeoanthropologists led by Mary Leakey began to throw elephant dung at one another. One exuberant member of the party noticed the prints only after falling on them.

But perhaps the most famous footprint, at least the one pressed most deeply in the Western imagination, was never really made at all:

> It happen'd one Day about Noon going towards my Boat, I was exceedingly surpriz'd with the Print of a Man's naked Foot on the Shore, which was very plain to be seen in the Sand: I stood like one Thunder-struck, or as if I had seen an Apparition; I listen'd, I looked round me; I could hear nothing, nor see any thing . . .

there was exactly the very Print of a Foot, Toes, Heel, and every Part of a Foot; how it came thither, I knew not, nor could in the least imagine.

This discovery of a single footprint is the most iconic moment in Daniel Defoe's *Robinson Crusoe*, published in 1719 and sometimes referred to as the first modern novel. Robert Louis Stevenson considered it one of four emblematic scenes in literature that, more than any other, have been 'printed on the mind's eye forever'. Friday's impossible mark – how can there be only one, isolated in the middle of an otherwise pristine beach? – has Crusoe spooked. After enduring the solitude of his deserted island, he suddenly sees hints of human presence everywhere, 'mistaking every bush and tree, and fancying every stump at a distance to be a man'.

The discoveries of Friday's footprint and the footprints of early humans have such a vivid claim on our imagination because we have all lived a version of it at some point: the sudden feeling of being accompanied by an unseen other. Although you are alone, the air seems somehow closer, or an empty room is still thick with the presence of one only just departed. Someone or something has passed through already.

In the final part of *The Waste Land*, T. S. Eliot drew inspiration from accounts of the Shackleton expeditions to Antarctica, in which exhausted members of the party would hallucinate that there was always one more person present than could be counted. 'When I look ahead up the white road,' complains one of the poem's many disembodied voices, 'there is always another one walking beside you.' It was recently suggested that the Laetoli footprints do not represent a pair walking side by side, as was originally thought, but sets of individual prints made at different times. New high-resolution photographic techniques have revealed a third set of toeprints obscured by the other two. The third walker seems to have favoured their left foot over their right and was perhaps carrying an injury. Wherever they were going, they did not come back by the same route: there are no footprints to mark the return journey.

AS ONE SET OF FOOTPRINTS stepped out of the past, another kind stepped into the future. In May 2013, the same month that the Happisburgh footprints were uncovered, climate scientists at the Mauna Loa Observatory, in Hawaii, announced that atmospheric carbon dioxide levels had reached 400 parts per million (ppm) for the first time in all human history.

For the past eight hundred millennia, since the Happisburgh footprints were pressed in the mud to the middle of the nineteenth century, atmospheric CO_2 has oscillated between 180 and 280 ppm as the planet moved between freezing ice ages and warm interglacial periods. The last time concentrations were higher than 280 ppm was during the mid-Pliocene, three and a half million years ago, when the Laetoli footprints were made and our earliest ancestors were only just beginning to diverge from apes. It was in many ways a world we would recognize: the continents were essentially in the same positions they occupy now, inhabited for the most part by similar species of plants and animals, while the same kinds of fish swam in the oceans between them. But the seas themselves were tens of metres above where they are today, and the global mean temperature was around 3 degrees Celsius higher.

If the Pliocene world resembles the one we know, it also, potentially, foreshadows what our world will become. Some scientists look to the mid-Pliocene as a form of 'palaeo-laboratory' to better understand the difficult and dangerous world we will live in if the planet continues to heat up. Global mean temperatures are already 1 degree Celsius higher than they were in 1850, and we could reach 1.5 degrees Celsius by the middle of this century, leaving us poised on the threshold of a world that is radically different from the one modern humans evolved in. Already, drought and flood, wildfires and storms, have become more common in many parts of the world, with fatal consequences, but beyond an increase of 1.5 degrees Celsius, we face the prospect of having to rapidly learn how to live on a planet that has become profoundly alien: where crops do not grow as they once did, equatorial cities have become uninhabitable, and low-lying islands and nations sink beneath the sea. Perhaps a fifth of the ecosystems on Earth will undergo some kind of fundamental change if we cross this threshold, but far more worryingly, it would trigger the irreversible thawing of Arctic permafrost, releasing catastrophic levels of greenhouse gases and guaranteeing a return to a Pliocene-like climate within several centuries.

The new Pliocene is not a given, yet. We can still determine a different future. Even so, signs of the changes we have already wrought abound and will be evident to those who inhabit that future, however distant. Much of the carbon dioxide that rose from the furnaces of the industrial revolution and the exhausts of the very first combustion engines still circulates unseen above our heads, while the distinctive isotopes that come from burning fossil fuels are scattered like spores over the entire planet, layered in glaciers and lake sediments. Even if we were to stop using fossil fuels immediately, traces of the carbon we have produced would still linger for time out of mind. David Archer, a climate scientist at the University of Chicago, has estimated that as much as a third of the carbon derived from burning fossil fuels will remain in the atmosphere a thousand years from now. After ten thousand years, this declines to between 10 and 15 per cent, but the last of a long tail of around 7 per cent of anthropogenic carbon won't be weathered away for one hundred thousand years, long enough to delay future ice ages. Our carbon could influence the climate for the next half a million years.

The entire atmosphere now bears the marks of our passage, like a vast geochemical trace fossil of the journeys we have taken and the energy we have consumed. When the last residue of our carbon finally leaves the atmosphere, humanity will have lived and evolved through another four thousand generations. Language and communication will have altered beyond our ken; how people in the year 102,000 CE talk and think, what they consider to be art or music, may well be unrecognizable to us. What it means to be human may even have changed in ways we cannot imagine, but as that change unfolds and our descendants pull away from us, like the ghostly third figure in Eliot's poem, we will accompany them.

To produce the spike in atmospheric carbon measured by the Mauna Loa scientists, we have also made countless other deep marks, from the burrows we have dug in pursuit of fuel or minerals to the network of hard-wearing roads that carry them from pit to pump or factory. Our carbon traces won't be directly legible without specialist knowledge and equipment, but we can already read them in the form of more frequent and intense extreme-weather events. New landscapes shaped by climate change will silently bear witness. Droughts that parch the land or storms that flood it may produce trace fossils of their own, as ecosystems change or collapse altogether and rising seas make living in coastal cities untenable. A huge proportion of anthropogenic carbon isn't in the atmosphere at all but has been absorbed by the oceans, which are becoming progressively warmer and more acidic, with severe consequences for virtually everything that lives in or depends on them.

The moment I realized the uncanny coincidence of the discoveries in Happisburgh and Hawaii was both thrilling and appalling. In part, it was the curious intimacy of a connection across such a broad stretch of time. Like Crusoe, the Happisburgh prints give us 'Toes, Heel, and every Part of a Foot' — individual bodies that walked, and feared, and loved just as we do. But I wondered whether the 'footprint' we have left in the atmosphere will also inspire the same sense of recognition. Will future generations feel the past rush toward them, just as 850,000 years contracted to a distance of a few metres when the Happisburgh prints were discovered? Will they, like Crusoe, be alarmed by the realization that our presence still haunts their passage? The footprint has become one of the most widely recognized metaphors for human impact on the planet. Particularly in the West, we are urged to be mindful of how the way we live produces a deeper or shallower chemical imprint on the world's atmosphere. Our carbon footprint is a mark of how much we care (or don't) about the consequences of our actions. Sometimes the metaphor is literal, such as the famous exhortation that hikers 'take only photographs, leave only footprints'. But the suggestion that a footprint is ephemeral, a temporary impression soon wiped clear by wind or rain, masks the reality that our marks will endure for a very long time indeed. Our trace fossils will be inscribed in the planet's geological, chemical and evolutionary history, legible, in some cases, even to our most distant successors. Long after we are silent, they will speak of how life was lived in the late twentieth and early twenty-first centuries.

We can only speculate about who might notice them many years from now, if at all. Perhaps no one will be around to read our traces, but nonetheless we are, everywhere, constantly, and with the most astonishing profligacy, leaving a legacy that will endure for hundreds of thousands or even hundreds of millions of years to come. Like the Happisburgh prints, what seems most fleeting presages the most incredible leap in time. We are conjuring ourselves as ghosts who will haunt the very deep future.

I TEACH ENGLISH LITERATURE at the University of Edinburgh. In early 2013, only a few months before the Mauna Loa announcement and the discovery at Happisburgh, I began teaching a course concerned with writing about nature and place. Since then, once a week in the spring semester, my students and I gather in a small room around tables of lacquered blond wood, a surface that feels more like plastic than pine, and talk about the work of writers like Edward Thomas, Kathleen Jamie, and W. G. Sebald. One side of the room is all windows that open out onto a view of Salisbury Crags, a wave of fine-grained dolerite cliffs breaking at the foot of Arthur's Seat, the extinct volcano around which Edinburgh has wrapped itself for more than a thousand years.

My fascination with the idea of deep time began with teaching this course in the shadow of Salisbury Crags. Like the hub of an enormous wheel, the crags are both an emblem and a point of orientation for the city of Edinburgh. The path that runs around the base of the cliffs offers spectacular views of the rampart of the Pentland Hills to the south, and to the north and west, the Georgian New Town, the Forth estuary, and beyond to the low hills of Fife. But the crags hold an even more special place in the history of time. In the eighteenth century, while Edinburgh was the hub of an extraordinary clamour of intellectual activity known as the Scottish Enlightenment, the area was a quarry. A gentleman farmer named James Hutton used the crags to demonstrate his theory that sedimentary rocks are gradually lifted into mountains by immense heat and pressure from below the ground. His discovery of a fist of reddish igneous rock closed around a paler chip of much older dolerite, now known as Hutton's Section, proved that molten rock intruded upon older sedimentary layers. Hutton's *Theory of the Earth*, published in 1788, was the first scientific work to imagine the immensely long timescales required to shape the planet.

Hutton's ideas set him at odds with geologists today: whereas he saw the world as a machine, endlessly running through cycles of sedimentation that lifted the earth and erosion that ground it down, modern geologists acknowledge that the planet is shaped by sudden events as well as predictable processes, that cataclysms – a sudden increase in volcanic activity, or a meteor strike — have as much influence as the orderly cycles Hutton identified. Rather, his legacy is the scope he gave others to think with. His real innovation was to fundamentally change how we look at the world around us. Such a vision, measured in grains of sand, required hundreds of millions of years, deep time beyond anything that had been envisaged before.

Hutton's section was one of the first places where deep time was imagined, but the phrase is not his. Curiously, it first appears in a reflection on how long good writing might last. 'All work is as seed sown', wrote the Scottish polymath Thomas Carlyle in 1832, in an essay on James Boswell's *The Life of Samuel Johnson* that speculated on the longevity of Johnson's writing. 'It grows and spreads and sows itself, and so, in endless palingenesis' – or rebirth – 'lives and works. Who shall compute what effects have been produced, and are still, and into deep Time, producing?' Just under 150 years later it was popularized by the American essayist John McPhee in *Basin and Range*, a book about the landscape of the south-western United States. But like the fingers of magma that fired his imagination, Hutton's vision of deep time intruded upon the minds of the poets and writers who followed him. We can read traces of Hutton's thinking in Tennyson's *In Memoriam* ('The hills are shadows, and they flow / From form to form'). In 'On the Sea', Keats fancies that the ocean 'gluts twice ten thousand Caverns', and in Shelley's 'Mont Blanc', the slow violence of glaciation becomes 'a flood of ruin', producing landscapes 'ghastly, and scarr'd, and riven'. For some, deep time replenished a sense of mystery depleted by the loosening bonds of religious faith. 'How little do we know of the business of the earth', wrote Edward Thomas, 'not to speak of the universe; of time, not to speak of eternity'. Without Hutton's insight into the great age of the planet, Charles Darwin would not have had the scope to conceive his theory of evolution. From the perspective of very deep time, the most intractable rocks appear as fragile as eggshells, as free flowing as water.

Treating the planet as a succession of sinks and taps, as we have, has kept us focused on the present, concealing the fact that we also inhabit this flow. Earth's long pulse shapes the arc of our lives, but to see this poses a tremendous challenge to our everyday imaginations. For the most part, deep time is 'the strange sleep', which, according to Shelley, 'wraps all in its own deep eternity'.

One day in November 1944, standing on a 'bare ocean-moulded hill' in the chalk uplands of Dorset, the Irish writer John Stewart Collis sought to peer through the veil. 'I pressed my mind back through the bottomless abysses of time', he wrote later. The effort is beyond him, but it summons a memory of when time did, briefly, stand revealed:

Once, in the middle of the Atlantic, looking at the horizon, I tried to imagine the space beyond it. For a second I had a true glimpse of that space, and of the space beyond that space. And perhaps for as much as a second now, I saw the reality of a hundred million years.

In the ocean's immensity, the truly deep age of the earth flares for an instant with the force of a vision. In the rhetoric of ancient Greece, the term for this irruption of clarity was *enargeia*, and it described a speaker's capacity to peer beyond the present moment: Aristotle wrote that *enargeia* allowed an audience to 'see things occurring now, not hear of it as in the future'. What Collis saw as he sought to push his mind's eye beyond the grey horizon was the *enargeia* of deep time, rhyming the pitch and roll of the Atlantic with an uncanny tilt of the senses. The same vision is available to us, too, if we choose to look with patience and care, and by it we can catch, as Shelley did, 'gleams of a remoter world'.

Or perhaps not so remote. What *enargeia* reveals is not always easy to face – the poet Alice Oswald's translation of the term is 'bright unbearable reality'. Not long after the peak in May 2013, global atmospheric CO_2 dropped below 400 ppm, but this was only a brief reprieve. Allowing for fluctuations, today the level of CO_2 in the atmosphere is around 410 ppm, rising at around 2 ppm per year. Climate scientists at the Australian National University recently proposed that human activity is forcing changes to the Earth system 170 times faster than natural processes. By this queasy calculus, we will see ten thousand years of environmental change in fifty-eight years, less than a single lifetime.

Some geologists think that this staggering rate of change justifies the naming of a new phase in planetary history. For more than one hundred years the International Chronostratigraphic Chart, which lays down the sequence of geologic time, has culminated with the Holocene, the period of benign climate that began around 11,700 years ago with the end of the last ice age and coincided with the development of human society. But in 2009, the International Commission on Stratigraphy (ICS) charged a group of geologists, biologists, atmospheric chemists, polar and marine scientists, archaeologists and Earth scientists with establishing whether or not the chart should be updated to reflect the onset of a new unit of geological time: the Anthropocene, or the time of the human. The Anthropocene Working Group has focused its efforts on a search for evidence of wholesale change in the way the earth works as a system of interdependent geochemical, sedimentary and biological processes. For the evidence to be compelling, they determined, it must produce new and distinct layers in the stratigraphic record. The group explored human-mediated acceleration in the rates of erosion and sedimentation, disturbances to the major chemical cycles (carbon, nitrogen, and phosphorus), the likelihood of significant changes in sea level, and the effect of human activity on the diversity and distribution of species across the globe. They examined the potential for synthetic materials, from artificial radionuclides produced by nuclear testing to plastic waste, to leave an identifiable signal in the strata. Many of these changes and signals, they concluded, are not only present and observable now but also effectively a permanent part of the archaeological and stratigraphic record.

When determining the boundaries of geologic time, stratigraphers search for sites where evidence of the shift from one geological age to another glitters in the dark of deep time. Such boundary sites are sometimes referred to as 'golden spikes', and marked by a bronze plaque hammered into the rock. But what the Anthropocene Working Group sought was the flash of *enargeia*: not the residues of worlds past, but the difficult brightness of a new world arriving. Geology is a cautious discipline: many who practise it feel that the process for introducing a new entry in the International Chronostratigraphic Chart ought to be as patient as forming a novel layer in the strata. But in 2016, at the International Geological Congress in Cape Town, the members of the ICS voted nearly unanimously that the Anthropocene was a stratigraphic reality and that it coincided with the eruption of technological innovation and material consumption in the middle of the twentieth century. The AWG is currently working on a proposal to formalize the Anthropocene as a new unit in geologic time.

Hutton learned to read the deep past in the rocks he saw every day, and according to the Anthropocene Working Group we can now read something of the deep future in even the most ordinary artefacts. The evidence of the Anthropocene is all around us, inextricably woven into the way we live our lives. But to see it, we need to face the 'bright unbearable reality' of the world we have made.

IN MY CLASS, as the crags loom darkly outside our window, we busy ourselves with words on the page. For ten weeks, my students and I share ideas about what others have said about the natural world. We tour vicariously through Scottish moors and English woods, following in the footsteps, if only figuratively, of writers who have traced a river from source to sea or pursued a bird of prey across winter fields. Field trips aren't usually undertaken by literature students, but as if to acknowledge that ours is only a second-hand study of nature, at the end of the course we finally step outside the classroom. One Saturday morning in March we board a train to Dunbar, fifty kilometres east of Edinburgh, on the Lothian coast.

Our route from the train station to Barns Ness Lighthouse is only around twelve kilometres there and back, along a low, rocky shore. The walk begins by skirting the edge of the well-disciplined greens of the town golf course, following a thin ribbon of turf marked out for the benefit of walkers. The manicured lawns contrast strikingly with the disorderly flotsam piled up where the green gives way abruptly to pebbled beach. But as the last hole tapers away in a tangle of unruly grass, a much more complex scene starts to take shape.

It is, in truth, a rather functional landscape, pinned to a narrow strip of coastline by the grey barrier of the A1, the faint susurrus of distant traffic mingling with the sighing waves. At the far end of the beach that curls away from the golf course, a modern cement works, fed by a huge open-cut pit, overlooks a clutch of derelict nineteenth-century kilns, relics of the time when layers of coal and limestone were stripped and burned to provide quicklime for local farmers. The kilns, too dangerous to enter, are encircled by a chain-link fence and wreathed with warning signs. The whole scene sits on top of a limestone pavement, source of the materials cooked up in the kilns 150 years ago. Most of the fossils here are, like the Happisburgh footprints, trace fossils. Thousands of curved tubes, the tiny marks of long-extinct animals burrowing for shelter or food, are strewn across the pavement like pieces of macaroni. A large area is pockmarked with dozens of shallow basins, thought to mark the sites of individual trees that grew in a tropical Carboniferous forest when Scotland lay nearly at the equator. Some basins are filled with seaearth, a fossilized wetland soil in which you can still see the fine tracery of ancient roots.

As the naturalist Adam Nicolson has said, in geological terms northern Europe is a landscape in recovery, still reeling from the immense trauma of glaciation. Since the ice sheets melted away, the British Isles have been rising slowly through a process called isostatic uplift – rebounding into shape like a pillow relieved of the weight of a sleeper's head. Just as the mountains in the Highlands of Scotland, which once rose higher than the Himalayas, have been ground down to nubs, the town, the motorway, the lime kilns, and the cement works will weather away over time until virtually no trace remains. But before that erasure they will have marked the earth indelibly. The cement works are a reminder of the truly sublime quantities of concrete that we have produced, and the processes involved. Humans have been earth movers for thousands of years. It's thought that if all the evidence of human geomorphology to date were heaped together, the spoil would form a mountain range four thousand metres high, forty kilometres wide, and one hundred kilometres long. But by the end of the twenty-first century we will have shifted as much stone and sediment in 150 years, through mining, construction, and road building, as humans moved in the preceding five millennia. Every year we move around eighteen thousand times more rock than the Krakatoa eruption in 1883. Around half a trillion metric tonnes of concrete have so far been cast for human use, enough to spread a kilogram layer across every square metre of the earth's surface, half of it produced in the last twenty years.

A few miles to the south of the limestone pavement lies Torness Nuclear Power Station. In time to come, there will be nothing left of the installation itself except perhaps an irradiated patch of ground. But the waste it has produced, even in the thirty or so years since it opened, will leave a trail across the globe. Much of the uranium processed at Torness comes from Australia, from underground mines like Olympic Dam in South Australia or open-cast mines like Ranger in the Northern Territory – a vast crater, stepped like an Incan city, which has displaced tens of millions of tonnes of rock. At present, spent fuel from Torness is delivered to Sellafield in Cumbria, the largest nuclear facility in the UK, along with 80 per cent of all the country's high-level waste. Thousands of cubic metres of waste, accumulated in the first four decades since the plant opened in the 1950s, are still stored in huge open-air concrete storage ponds. Photographs leaked to the press in 2014 showed seagulls bathing in the water. In some of the oldest laboratories, now decommissioned, it is unclear exactly how much or what kind of deadly material they contain. Most of the waste received at Sellafield today is reprocessed, but a stubborn remainder of around 3 per cent is left over. In lieu of a more permanent solution, this is mixed with liquid glass at 1,200 degrees Celsius. When it cools, the mixture vitrifies, forming solid blocks of irradiated glass. Sellafield houses six thousand steel containers of vitrified waste like huge toxic sugar cubes. The bitter material within will be lethal for thousands of years – still harmful to people for whom we will be little more than a rumour.

Other, more banal materials on this shore possess the same astonishing reach through time. We all carry packed lunches with us on our trip, which include a lot of foil- or plastic-wrapped sandwiches. We diligently take all our rubbish away with us, until we can deposit it in the nearest bin. The majority of Edinburgh's household waste in fact ends up not far from this beach, in a landfill site lined with clay and plastic. Most modern landfills are constructed this way, creating an airtight and watertight seal to prevent toxic materials from leaching into the groundwater, effectively mummifying their contents. In the 1970s, an archaeologist called William Rathke became interested in what happens inside landfills. For the next twenty years he excavated sites around Tucson, Arizona, and reported finding forty-year-old hot dogs, twenty-five-year-old lettuce still in storefront condition, and – in the mid-1980s – an order of guacamole that looked ready to eat despite having been buried alongside a newspaper from 1967. If food can last for decades in mid-twentieth-century landfills, more durable materials like plastic and aluminum buried in modern landfill conditions will certainly retain recognizable forms for far longer.

Since the middle of the twentieth century we have produced enough aluminium, around five hundred million metric tonnes, to cover the whole of the United States in kitchen foil. The majority of the millions of tonnes of plastic that enter the ocean each year falls to the ocean floor, where it will be folded into the sediment as a layer in the geological strata, effectively a permanent addition – at least until the heat and pressure turn it back into oil or the section of seafloor is raised up and eroded, processes that will be measured in tens of millions of years. Even the contents of our sandwiches can tell a story. Sixty billion chickens are killed for human consumption each year; in the future, fossilized chicken bones will be present on every continent as a testimony to the intrusion of human appetites in the geological record. These most ordinary and familiar things, each with the potential to become a new fossil, bring the intimacies of the Anthropocene up close.

As we leave for our train back to Edinburgh, we turn away from this beach, but it will remember us.

FOOTPRINTS IS MY ATTEMPT to discover how we will be remembered by the very deep future. People have been modifying the land and changing ecosystems for thousands of years, but the alterations to the planet and the ever-more-durable materials we (mostly in the global North) have made since the industrial revolution have come with unprecedented speed and invention, and will leave long-lasting marks, beyond anything humans have produced before. In my search for future fossils, I look to the air, the oceans, and the rock, from a bubble of ice drawn from the heart of Antarctica to a tomb for radioactive waste deep beneath the Finnish bedrock. I examine the landscapes and objects that will endure the longest and the changes they will undergo: the processes that will transform a megacity into a thin layer of concrete, steel and glass in the strata; the future of the fifty million kilometres of roads that circle the planet and supply our cities with materials moved over vast distances; and the stories of those materials themselves, like the five trillion pieces of plastic waste already circulating in the world's oceans.

But it is also a search for what will be lost. As biodiversity declines, silence will itself be a signal, absence another kind of trace. Bleached coral reefs, like the one I saw in Australia, will be monuments to this loss, but so will marine dead zones such as the huge area of anoxic water I visited in the Baltic Sea. Ice cores represent an astonishing archive of past climates, including the changes introduced by human activity, but as the ice melts, part of this record will go with it, while the loss of ice will write a new story in the planetary archive. There are also dangerous and highly durable substances like nuclear waste that we hope will remain hidden and forgotten altogether. And for all the many marks we will leave that cannot be mistaken – the deep pits we gouged in the earth and the rich pockets of landfill that hoard our waste – we will also leave our impression on worlds we can't see. Microbial life is responsible for engineering virtually every key life process and chemical cycle, flooding the atmosphere with life-giving oxygen, but its role has been usurped. At the end of my journey I examine how our prints will linger in the cells of some of the smallest life-forms on earth.

To perceive future fossils means to see what the Anthropocene's bright unbearable reality reveals; to look at a city as a geologist might, and to approach the problem of making nuclear waste safe from the perspective of an engineer; to understand the chemical stories in a piece of plastic waste, and to listen to the silences that echo in collapsed ecosystems. But it has also sent me back, again and again, to the essential elements of what I talk about with my students: to narrative, myth, image, and metaphor. I want to discover the world we will leave behind, but also how we will appear to the people who may live in that world. It's an account of what will survive of us, and for that we need poets as much as we need palaeontologists. With stories we can see the world as it is and as it might be; art can help us imagine how close we are to the extraordinarily distant future.

We already know that the Anthropocene is a global story, but we don't need to go far to find evidence of it. Future fossils are all around us, in our homes, in our workplaces, and even in our bodies. So my journey began in Edinburgh, and while it also led me to some very faraway places, it returned periodically to the North Sea world in which I feel at home. Much of my search also took place while I held a visiting fellowship at a university in Sydney, about as far from Scotland as it is possible to go, and a sweltering contrast with the northern climes I am accustomed to. At times, I found I had to seek out particular places to better understand their role in shaping our future traces: to learn about how cities might become fossils, I visited Shanghai, a city of twenty-four million people that has sunk under its own massive weight by over two metres in less than a hundred years. But what struck me most forcefully was how ubiquitous future fossils are. Our present is saturated with things that will endure into the deep future. As you read this, you will also, in all likelihood, be surrounded by objects and materials that could contribute to making a trace fossil. Before you begin to take this journey with me, look up from the page and imagine how the things around you – the plastic casing of your laptop and its titanium innards or the coffee cup standing beside it – might remain, even just as an impression in stone, millions of years from now.

Future fossils are not just a distant prospect to be left to the patient care of geological processes or the curiosity of generations yet to be born. They touch our lives hundreds of times every day, and we can see in them, if we choose, not only who we are but also who we could be. We have already fundamentally altered the systems that support life on the planet, in ways that are deeply sobering. The most vulnerable will be the worst affected, and the full costs to future generations have yet to be calculated. Our future fossils are our legacy and therefore our opportunity to choose how we will be remembered. They will record whether we carried on heedlessly despite the dangers we knew to lie ahead, or whether we cared enough to change our course. Our footprints will reveal how we lived to anyone still around to discover them, hinting at the things we cherished or neglected, the journeys we made and the direction we chose to take.

ONE

THE INSATIABLE ROAD

It was billed as a once-in-a-lifetime opportunity: the chance to walk Scotland's newest stretch of road, spanning the banks of the Firth of Forth.

Since 1964, all road traffic across the estuary had been borne by the Forth Road Bridge, a burden comprising hundreds of millions of journeys north and south. The old bridge had begun to show the strain, though, and so a new one was commissioned. It took six years to complete. My family had followed its unhurried construction as the deck inched over the water and the web of cables slowly knitted together. From the beach near our house we could see the rising towers nudge above the hill between Edinburgh and South Queensferry, where the bridge was being built. Whenever we drove west from the city, my children would point out changes in its shape and size. Now it was finally ready to open, and to celebrate, a ballot had selected fifty thousand people to walk its 2.7-kilometre span across the estuary. We were lucky enough to be among them, and so on a golden Saturday in September, we set off to make on foot a journey that would thereafter be possible only at fifty miles per hour.

We met the bus that would take us the eight kilometres or so to South Queensferry in an industrial park on the outskirts of Edinburgh, and the new bridge rose into view as we travelled west along the estuary. At a distance, the Queensferry Crossing is a miracle of light and air, held in place by gleaming white threads strung from three spindle-like towers. The cables that lace it together resemble strings on the soundboards of a series of upended pianos, and the deck rises and dips like the harmonic curve in the neck of a harp. 'How could mere toil align thy choiring strings!' wrote Hart Crane of Brooklyn's famous bridge. I wondered what enchanted music a strong North Sea wind might make as it barrelled down the estuary.

Our bus pulled into the empty multi-lane motorway just before the bridge's southern rampart rose over the water, and we joined the crowds strolling north towards Fife. As the asphalt crunched under my feet, the airy impression I'd had from a distance resolved into something much weightier. The white cables, which had looked so slender, were thicker than my body. Viewed obliquely, they seemed to bond into a single white wall. The road surface was hard and unyielding, and fist-like rivets bulged from every knuckled stanchion and guardrail. The lightness, rather, was in me. I felt giddy walking on a surface never intended for foot traffic; it was as if in stepping out over the water we had also stepped into a wholly different relationship with the space around us. Uniquely, the bridge's textures were there to be sampled: the cables' bone-white smoothness, the glaucous sheen on the barriers between the carriageways, the road's coarse grain. There was a thrill in the air, a mood of trespass. In reality, the event was subject to airport-levels of organization and stricture; bag-searched and photo-ID'd before we arrived at the bridge, we had firm instructions not to take longer than an hour or we would risk missing our return bus. But for a brief, beguiling moment, it felt like we were reclaiming the road.

Really, we have conceded so much. Most of us live and wander only where road networks permit us to, creeping along their edges and lulled into deafness by their constant roar. Man 'sets his house upon the road', lamented Ralph Waldo Emerson in 1849, and every day the human race goes forth and cuts a path for him to follow. But on this journey we could roam where we pleased, no longer confined to the curb; rather than the growl and whine of engines, the sounds were light and textured, of voices, laughter and the soft crump of hundreds of footsteps. This new road, built to withstand more than twenty million motorized journeys every year, seemed more like a road out of the pre-industrial past, a pilgrim route made by tramping feet. Or perhaps a vision of roads to come, when the oil is gone and the engines are silent.

At the base of each tower, a large sign listed the facts and figures of its construction. The bridge, which at a distance seemed to float above the water, was bound to the earth by superlatives. One hundred and fifty thousand tonnes of concrete and thirty-five thousand tonnes of Chinese steel, shipped to Rosyth from Shanghai shipyards, went into its construction, and thirty-seven thousand kilometres of cables, only just short of enough to circle the Earth at the equator. Laying the foundations for the south tower involved the longest ever continuous underwater pour of concrete: nearly seventeen thousand cubic metres, tipped night and day for fifteen days into the rock beneath the river. Excavating the site for the new network of roads that would connect to the bridge had revealed the remains of a Mesolithic sunken-floored house, the oldest dwelling ever discovered in Scotland. The remains of a clutch of post holes, now just shadows in the earth, along with charred hazelnut shells and fragments of burnt bone had survived in the mud for perhaps as long as eleven thousand years. But the aggregate seam of concrete pressed into the bedrock beneath the south tower, crushed Scottish granite or English limestone mixed with sand from India or China, will have a far more long-lasting presence, posing a riddle for future geologists to puzzle over.

A rude honking from the river barged through the chatter of voices as a container ship passed beneath us, sounding its horn in acknowledgment as it followed its own watery road.

At the bridge's northern end, a small crowd had gathered around a knot of photographers. Scotland's first minister was giving an interview, and we hovered for a chance to take our children's picture with her. As they grinned for the camera, I looked to where the road flowed north through a coil of raised carriageways and sunken slip roads. Maybe a hundred metres away, a huge dolerite cliff loomed along the road's eastern side. Engineers who built the first rail crossing over the estuary in the 1880s had punched their way through the gently rolling landscape, exposing to the air stone that had not felt wind or rain for millennia. They had cleaved their way through this mound of rock as if they were splitting a skull. If I had been passing over the bridge by car, I would have had barely seconds to notice it, lulled by the blur of asphalt slipping like grey silk under the car's nose, perhaps registering the weight of rock as no more than a shadow in my peripheral vision. But free as I was to stand and stare, the exposed stone seemed to catch me up out of the present and draw me in, and through, and down into the memory of a younger earth.

At my back, that huge bolus of concrete slumbered beneath the river, curled around the base of the south tower like a dragon around a hoard of gold. As I gazed at the cutting, the bridge ceased to be a connection between the opposite banks of the river; it was, for an instant, poised between moments in time that flew beyond my imagination.

One million years from now, the bridge's thin towers, its choir of shining cables and elegantly curving deck, will be long gone. The surface of the road will be washed away. But even as the eroding forces of weather and time take their toll, grinding down the cliff and filling the engineers' clefts with sediment, the concrete foundation and the rock cutting will still be legible, written into the earth like speech marks around a lost quotation, bearing witness that here, once upon a time, a road crossed a river that will itself long since have vanished.

IT IS SAID that the world's longest road is the Pan-American Highway. Really a dense network of interstate roads that wend and purl through seventeen countries, the highway runs unbroken from Alaska to the foot of Argentina, except for a 160-kilometre belt of rainforest between Central and South America. Its northernmost point is Prudhoe Bay, on Alaska's Beaufort Sea, the site of the largest oil field in the United States. Barry Lopez writes that the thousands of oil-production wells that punctuate the bay make it seem more like a portion of

West Texas transplanted to the Arctic tundra. From here, the road follows a gentle arc through Alaska's Brooks Range to Fairbanks, then south into the Yukon, where it bends east around the northern tip of the Canadian Rockies and surges over the plains of Alberta, skirting the Athabasca tar sands, to Edmonton, where the road forks. One tine heads south and east, reaching the edge of the Great Lakes before doubling back along the rim of Lake Superior to Minneapolis, followed by Des Moines, and the Great Plains cities in Kansas and Oklahoma, and on to Dallas, past the sprawling oil fields of East Texas. The other takes a line south and west to Calgary, cuts through the Blackfeet, Flathead, and Crow reservations of Montana, then to Wyoming and the new shale plays in Western Colorado to Denver. Past Albuquerque the road begins to curve east, dipping under the wing of the West Texas oil fields, to meet and rejoin itself at San Antonio.

Jack Kerouac's last trip in *On the Road* followed some of this western fork, from Denver to Mexico, as if it were an enchanted journey to reach a fabled city. Kerouac declared it the most fabulous road of all: miles and miles of 'the magic *south*'. With Dean Moriarty, he burned through a thousand Texas miles, past a seemingly endless succession of gas stations, to San Antonio, then arced south to Monterrey through a gap in snow-topped mountains, crossing the swamps around Montemorelos and the desert plain, to where, he said, every road seemed to point: Mexico City.

Kerouac's journey ended here, in spring 1950, in a sump of tropical fever that sent him limping back to New York. But the road that fuelled his imagination continues to push south. From Mexico City, it pours like sand through the hourglass waist of Central America to Panama, where it crosses the canal on the Centennial Bridge. Two hundred and sixty kilometres farther south, the road breaks, briefly and for the only time in its enormous span, frustrated by a barrier of rainforest and mountains called the Darién Gap, where Keats once imagined Cortez standing captivated by his first glimpse of the Pacific Ocean. It resumes in Colombia, winding through the Ecuadorian highlands to Quito, west of the oil fields of Lago Agrio and Pungarayacu, and skirting the edge of the Amazon rainforest. Here it tucks in behind the rampart of the Andes and follows the Pacific coast, past more oil fields in the waters lapping around Lima, to Valparaíso in Chile, where it abruptly strikes east along Route 60 (passing through the three-thousand-metre Cristo Redentor tunnel driven through the roots of the Andes) to the baroque esplanades of Buenos Aires.

The final leg of the journey hugs the Atlantic coast all the way to Tierra del Fuego. Bruce Chatwin describes travelling along this section of the road in *In Patagonia*, where he notes that the conquistadors named the region for the billowing domestic fires of the Fuegian Indians. Magellan called it the Land of Smoke (Tierra del Humo), but, Chatwin claims, the Holy Roman Emperor Charles V ordered that it be renamed, reasoning that no smoke exists without fire. Chatwin's journey took him through the Land of Fire along the final stretch of Argentina's National Route 3, lit by the flares of oil rigs in the southern Atlantic rather than Magellan's fires, to its terminus at Ushuaia, the world's southernmost town, forty-eight thousand kilometres from where it began.

Modern roads connect the world we have made. It's

estimated that there are more than fifty million kilometres of roads worldwide, at least a third of which are sealed. That is enough paved surface to loop around the planet thirteen hundred times. China alone has more than four million kilometres of paved roads. The story our future fossils will tell is, in some respects, determined by this network. Many trace fossils are marks of passage, detailing where a creature passed through long ago. Although made by our machines rather than our bodies, roads will be as telling as any footprint in this regard. It's a tale of massive displacements, of huge quantities of materials drawn from one place and laid to rest, like the bulb of concrete planted at the base of the Queensferry Crossing's south tower, somewhere far away. But it's also a story of the places that make the modern world possible – places excavated for their resources and left to ruin – that might seem very distant to those of us in the comfortable West, but to which we are intimately connected. And it's about what flows through boreholes, pipes and engines and drives our need to keep extending the road: oil. 'Oil is a fairy tale', writes Ryszard Kapuściński. But like every fairy tale, Kapuściński cautions, oil is also a lie. It promises release, but in reality oil sticks us to the shadow places. To know this story, we need to know not only what will become of the roads themselves but also what they connect us to. Sealed asphalt and concrete may not receive a footprint, but nonetheless the road will be a reliable source of future fossils.

FIRST, THOUGH, we need to contend with a problem of perspective.

Roads furnish our imaginations with images of freedom. Journeys like Kerouac's have come to stand for a sense of unimpeded progress and self-discovery, an open horizon connoting limitless possibility. Roads conjure what it feels like to be modern. They open up the world for us, but, as Emerson realized, they also dictate the direction we take. Roads accompany us for so much of our lives – how much time do any of us spend more than a hundred metres from a road, or out of earshot of their whispering voices? – and yet we have somehow trained ourselves not to really notice them at all.

In 1983, *Vanity Fair* commissioned the artist David Hockney to illustrate a story about the road trips Vladimir Nabokov undertook when he was writing *Lolita*. As he researched and wrote the novel in the 1950s, Nabokov crisscrossed the United States, driven always by his wife, Vera, in a 240,000-kilometre tapestry that stitched the east coast to the west. Hockney's own road trip began with a journey through the Mojave Desert during an April storm. His subject was proving elusive, and the weather didn't help. But the next morning, after some further desultory driving and photographing, Hockney proposed to his driver that they might find something promising at an intersection they had passed the day before. It took some time, but eventually they found it, and from the images he took over the next eight days Hockney would compose one of the twentieth century's most iconic images of the road.

Pearblossom Hwy., 11–18th April 1986, #2 is a trap for the viewer's perspective. A collage assembled from hundreds of individual photos, it depicts an unremarkable stretch of desert road where the Pearblossom Highway crosses California's

Route 138. The road tapers into the distance from a deep wedge in the foreground. Two parallel lines of thick yellow paint mark the division between lanes, running from the bottom edge of the picture to its midpoint, where it is bisected by the horizon of the blue, rippling Angeles Crest Mountains in the distance. Four road signs, yellow, green and red, track the right-hand side of the road towards its vanishing point; the scrub on either side is dotted with spiny Joshua trees and littered with discarded bottles, cans and cigarette packets. The upper portion of the painting is almost entirely taken up with the California desert sky, a big naive strip like the bar of blue a child might paint. The elements are simple and nondescript: asphalt, road signs, trees, mountains, sky. But the whole is dizzyingly precise. Every single image is taken in close-up and often head on (Hockney used a ladder to shoot the stop signs and to look directly down on each item of rubbish). Wherever it travels, the eye is arrested by detail; each crack in the painted road markings and flash of sunlight caught in the wrinkles of the crushed Pepsi can is intimately and immediately present.

Roads can do something odd to our sense of time and space. When we travel by road, we often do so in a kind of reverie. As a child, I would fill the boredom of long car journeys by imagining I was watching myself running alongside our car and leaping impossibly over the roadside clutter. It was a fantasy of perfect, unimpeded motion. Now, as I settle behind the wheel, other times and places – unresolved problems, anticipation of my destination, or unspooling threads of memories – fill my thoughts, and I find I can travel without really registering the world around me.

Seamus Heaney called this the 'trance of driving' and

composed his poems under its spell, beating out their measure on the steering wheel. Many roads are laid out to suppress our sense of place. High verges built to muffle traffic noise also obscure our view of what lies beyond the road; the bland grey crash barriers barely register. The white lines progress relentlessly towards the horizon. For Joan Didion, driving on the freeway around Los Angeles involves a form of concentration so distilled as to become narcotic, 'a rapture-of-the-freeway'. The lulling rumble of motor travel, the incantation of road signs marking the passage of miles, the low rush of our own slipstream – all can conspire to lift us out of time. In such moments we are, in a way, perfected. 'The mind goes clean', writes Didion. 'The rhythm takes over'.

The history of modern road travel is the pursuit of the perfect road, the most frictionless way possible. A smooth passage through space is perhaps the fundamental promise of modern life, like a spell cast to release us from the heaviness pinning us to the earth.

This transmutation began in the nineteenth century with the railways. In the 1830s, mechanized transport increased the speed of travel by stagecoach by a factor of three, and in doing so put travellers by rail into a fundamentally different relationship with time and space. An 1839 article in the *Quarterly Review* remarked with breathless enthusiasm that under the influence of rail travel, the world would 'shrivel in size until it became not much bigger than one immense city, and yet by a sort of miracle, every man's field would be found not only *where* it always was, but as *large* as ever it was!' Decades before Lewis Carroll subjected Alice to an enchanted regimen of shrinking and expanding, the railway had changed the world into a wonderland that mimicked the processes of geological time, contracting the planet's greatest rivers to streams, its lakes to mere ponds.

With the invention of the motorcar, roads were built to mimic the perfection of the railway. Whereas the old highways had to adapt to the contours of the landscape, contorting to accommodate each immovable peak and valley, the railway simply drove its tunnels and cuttings through the earth. Twentieth-century roads were built to the same technical standards, forcing the land to submit to the demand for frictionless movement. The historian Wolfgang Schivelbusch notes that the mechanization of travel induced a corresponding mechanization of travellers' sense of their place in the world. The velocity of rail travel – as much as forty miles per hour in the 1830s – destroyed the depth perception that defined pre-industrial place-consciousness. While objects at a distance could be observed in a newly panoramic vision, their aspect changing rapidly as the train sped past, the foreground was lost to a blur of indistinguishable shape and colour. Pre-industrial travellers, moving on foot or drawn by animals, would be immersed in their immediate environment, but after the railway most travellers would feel that they no longer occupied the same space as the objects they saw out of the window.

This detachment is the common experience of the road today. As we travel on it, we are caught up and carried away; inured to our surroundings, bounded by steel and glass, we are absorbed by infrastructure, numbed by vibration. As we observe the world through the cinema screen of the windscreen, our minds travel elsewhere while our bodies journey to their destination. Mechanical travel blunts our sense of the world. Emerson said the railway fed travellers' egocentrism, reinforcing the impression that 'whilst the world is a spectacle, something in himself is stable'. But in *Pearblossom Hwy. #2*, every intimate closeup detonates our sense of stability. Hockney's collage stitches us back into the scene. The dulling patterns that remove us from the world – the regular pulse of road signs marking distance reminding us that what matters is not where we are but where we will be, and the white ticker-tape road markings clicking by endlessly insisting that 'here' is in fact always several metres ahead – are replaced by hundreds of single moments. As if to remedy the enchantment of the perfect road, *Pearblossom Hwy. #2* returns us to the here and now – or, rather, to a here composed of countless 'nows'. The road, it seems to suggest, is a charm that must be broken.

SEVERAL MONTHS after I walked with my family over the new bridge, I set off early one Sunday morning to cycle over the old one. It was one of those immaculate November days, lit by the mica-sparkle of frost on quiet streets. The air tasted rich and golden. The only sound was the quick, dry chatter of brown leaves under my wheels and, once, the indignation of a skein of geese nagging one another across the tall blue sky.

The original Forth Road Bridge, the one built in the early 1960s, had been closed to cars since the opening of the new crossing. Only buses used it now, as well as cyclists and pedestrians, but these were infrequent. I'd seen the old bridge when driving over the new one or standing on the shore, and

it seemed abandoned. I wanted to stand on it free of the surging traffic it had once borne. Doing so, I felt, might give me
a sense of what will happen to roads themselves after we are
no longer around to travel on them.

As I cycled through South Queensferry, the cobbled street
telegraphed each bump through my front wheel. At the far
end of the high street, the road passed under the base of the
old bridge, its concrete flanks stained by half a century of
Scottish weather. To the left, a concrete path wound up the
bank to the deck. As I rose onto the bridge, the low sun cast a
brittle silver road on the water. A scatter of small fishing boats
nodded mid-river, and the cables of the new bridge shone like
a flotilla of sails about to depart. In the distance, I could see
Blackness Castle on the southern shore, a fifteenth-century
garrison known locally as 'the ship that never sailed' because
its tapered fortifications point prow-like over the estuary. A
single tanker was making its way slowly out to sea. Snow soft-
ened the Ochil Hills away to the northwest, and behind the
lower hills that ran down to the river to the east, the cooling towers of an ethylene plant at Mossmorran sent up an
immense gout of steam into the cloudless sky. Periodically,
the plant's operators burn off excess gas when the industrial
processes need to be restarted. The Mossmorran flare can
burn for days; when it does, I can see it lighting the sky from
my bedroom window. The most recent flaring had been only
a few weeks ago, blazing day and night like the eye of Sauron.

I have friends who live on a street near the old bridge and
have spent a lot of time in its shadow. The traffic noise used
to be constant, at times as thick as the haar, the sea fog that
sometimes rolls in here from the North Sea. But now all that

rush and clamour had evaporated into an eerie quiet. The wind was still, but even so, the sound of the thin line of cars on the Queensferry Crossing was swallowed up by the space between new and old bridges.

Whereas walking across the new bridge I had felt the lightness of a new beginning, crossing the old one, emptied like this, felt like an elegy. 'The end of the road' is often used figuratively to describe our sense of things coming to a close, or of terminal intent, but I think we rarely consider the end of roads themselves. The poet Edward Thomas knew this when he wrote, in 1911, that 'much has been written of travel, far less of the road'. But here was a road poised as if on the cusp of disuse, before the cracks began to appear and weeds embraced its towers.

At the bridge's end I passed through a deep cutting in the hillside to allow the road through. Its sides were decked cheerily with yellow gorse, but beneath this and a dusting of green moss the exposed rock glowered redly. I was reminded of Roy Fisher's beautiful poem 'Staffordshire Red', about the enchanted experience of driving through a rock cutting. Surprised by a turn in the road that leads directly through a sandstone cliff in the English Midlands, the poet finds himself, for a moment, plunged into a primeval landscape of dripping ferns and green light. Before he knows it, the road drops him back into the mild Midlands landscape, revealing the portal to be no more than a nondescript clump of trees. And yet he feels, he says, somehow altered, compelled to follow the road in a wide arc around the county until it returns to the cutting and pours him again down 'the savage cut in the red ridge', to feel once more the 'brush-flick of energy' touched off by this fleeting contact with an ageless mystery lying in wait among the ferns and moss.

I wheeled my bicycle beyond the cutting, to the knot of slip roads that carried traffic onto and off the bridges. Here, for around a hundred metres or so, the roads approaching both bridges ran in parallel. As I stood on the entrance to the stilled old bridge, I could see a steady line of cars moving onto the new one, and for a moment it was as if I were standing not on the old road but on a prophecy of that new road's future.

One day – whether because the exhaustion of fossil fuel reserves forces us to live within a more narrow compass or simply because humanity will, inevitably, no longer be around to use them – the roads that connect our towns and cities will be abandoned. The plant life we cut back to their edges will creep unchecked; their surfaces will split and rupture. Time will bring even the bold towers of this bridge down low. Although hard-wearing, most of it will be broken up and eroded. Persistent roots will worry away at their surfaces, and rains will wash them away. Some fragments, though, will be preserved as hints of the former whole. Like the oldest paved road in the world, a four-and-a-half-thousand-year-old stretch discovered near Cairo in the early 1990s, short sections will be buried beneath gathering sands, submerged by rising sea levels, or covered by land slips. Subject to unimaginable pressures that will warp and compress it, the hard-core base and asphalt surface will nonetheless be evident in the strata; and if, millions of years from now, the forces pressing one of these sections down into the earth are reversed, eventually the fossil road will be raised into the air like a new bridge. Embedded within this new cliff or mountainside, it will be a curious anomaly: a layer of rock that may have originated thousands of kilometres away from its resting place and a clue to the grey networks that once wrapped around the planet.

Tunnels have an even greater preservation potential, such as the twenty-five-kilometre Lærdal Tunnel in Norway – which is so long that it takes twenty minutes to travel through, and was built with three vast subterranean chambers, like the halls of mountain kings, each one lit to simulate sunrise to prevent drivers from falling asleep. The only real threats to their persistence into the deep future are earthquakes. The surface road network may yield only short fragments, each one only a kilometre or so in length. Perhaps less than 1 per cent of the Pan-American Highway will persist long enough to leave a fossil. Two hundred thousand years ago the Laurentide ice sheet reached as far south as Missouri – a new ice age would wipe away its entire northern section, while the weathering of the Andes will also erase stretches that pass through high latitudes. But the three kilometres that pass through the Cristo Redentor Tunnel would be protected, and tunnels like Lærdal and the Zhongnanshan Tunnel beneath China's Qinling Mountains could preserve stretches of fossil roads twenty-five kilometres long, complete with curbstones, road signs, lighting, and painted road markings.

I turned back to the empty bridge and began to cycle home. As I was halfway over the river, a bus roared past. The slumbering deck trembled briefly, then returned to sleep.

ONCE UPON A TIME, writes the Nigerian novelist Ben Okri, a giant called the King of the Road lived in the forest. But as the forest shrank due to the people's greed, he left and became the roads that the people travelled. He was a tyrant, with an appetite that could not be sated and the ability 'to be in a hundred places at the same time'. Travellers left sacrifices for safe passage, but still, the immense appetite of the King of the Road exhausted the land, and famine arrived. The sacrifices stopped, and, enraged, the famished king began to attack the living and the dead. To mollify him, the people assembled an enormous offering, enough to feed an entire village. They delivered it to the King of the Road, who swallowed it in a single mouthful and then proceeded to eat the delegation who had delivered it to him.

When a second delegation met the same fate, the desperate people decided to kill the king. They collected poisons from every corner of the earth and added them to a lavish meal of fish, bushmeat, yams and cassava. This time, the ravenous king turned on the delegation first, then ate the feast they had brought in one swallow.

After this meal the King of the Road lay down, and his stomach began to ache. To quell the pain, he ate everything he could lay his hands on: rocks, sand, even the earth itself. Finally, the king turned on himself, consuming his own body until only his insatiable stomach remained. Rain fell for seven days and washed the king's stomach into the earth. When the rain stopped, he was nowhere to be seen, but the people could hear his stomach growling beneath their feet.

'The King of the Road had become part of all the roads in this world', Okri writes in his novel *The Famished Road*. 'He is still hungry, and he will always be hungry'.

The road is insatiable. Paved roads link together all the most fundamental and long-lasting changes we have made to the planet's surface, from the deepest mines to the largest megacities; by them, we serve our addictions to finite resources. In the deep future our cities will be enormous sinks of countless future fossils, but virtually every last one will have originated far away and been transported to its new, distant resting place by roads. Each of the trillions of individual pieces of plastic in the world's oceans reached the coast via a series of journeys down highways, from oil field to hand. Roads themselves create huge quantities of synthetic particles abraded from tyres and washed into seas and rivers, where they will eventually settle on the seafloor, to be sealed under a layer of mud. Burning fossil fuels has coated the surface of the planet in a thin layer of fly ash. These tiny carbonate particles have no natural sources and are so widespread across the globe in lake sediments and ice cores that they rival nuclear fallout as the primary signature of the Anthropocene. It's thought that humans have modified more than half the planet's land surface in some form or another. Roads open up remote regions for exploitation, linking them to urban or industrialized centres. Gaia Vince notes that every road driven through the Amazon rainforest is pursued by a 'halo of deforestation' fifty metres wide, leading to more landslides and erosion and contributing to an acceleration in the cycling of sedimentary materials around the planet. Humans now move more sediment on an annual basis than all the world's rivers combined, around forty-five gigatonnes, increasing the likelihood that traces of us, including the roads themselves, will be buried and preserved as future fossils.

Underpinning all this is sand, the main ingredient in

concrete and asphalt. Global demand for sand is exceeded only by the demand for water. Around forty billion tonnes are used annually in construction and road building, as well as in the manufacture of window glass, smartphone screens, silicate solar panels and cosmetics. It is a key ingredient in metal foundries and fracking for shale oil and gas, and also used in the creation of artificial land. Singapore has used imported sand to add 130 square kilometres to its landmass in the past forty years, and when completed, the Palm Islands complex in Dubai (including an archipelago of islands shaped like a map of the world) will have used over three gigatonnes of sand, equivalent to the weight of nearly eight Great Walls of China. Despite its abundance, desert sand is too fine for commercial use; instead, we're dependent on the planet's capacity to weather enough coarse sand from the backs of mountains and the sides of hills, and global demand is outstripping geological processes. The King of the Road is hungry still.

Okri's Nigerian fable of the insatiable road sent me to the pictures of the Canadian photographer Edward Burtynsky. Since the 1970s, Burtynsky has photographed manufactured landscapes – quarries, salt pans, railway cuttings – in pursuit of what he calls the 'residual thing', the trace of us and our demand for raw materials that will linger long after the landscape has been relinquished by humanity. Often his subjects, far from urban centres, are what the environmental philosopher Val Plumwood has called shadow places: unacknowledged places, those that are unseen and unthought of, but feed our desire for minerals or energy.

Burtynsky makes his images on an epic scale in every sense, often taking them from a great height with the help of cranes, helicopters, or drones. Distance alchemizes the landscape, producing what Burtynsky calls 'mythic space'. From above, the scenes often resolve into patterns, revealing unnoticed geometries like abstract art, in which human figures are absent or reduced to tiny grains of colour (as he often works in industrial areas, the figures in the pictures usually wear yellow safety vests). The effect is detached, but not affectless. Burtynsky's landscapes are emptied of people but full of human presence. What we see reflected back is ourselves, or rather our shadow selves, our hungry selves, which have gouged, cut, blasted, shaped, and hoarded the earth until it looks back at us like a face in a mirror.

Roads have played a significant part in shaping Burtynsky's vision. His panoramic sense of landscape, he has said, emerged during long childhood journeys across Canada, watching 'the endless country go by'. As a young photographer searching for an aesthetic, he took a two-week journey alone around the United States. A wrong turn in Pennsylvania led him to a coal mining town called Frackville, where the landscape so arrested him that all he could do was stop the car and stare. Wherever he looked, in any direction, there was nothing in the land that had not been made by human industry. Mounds of coal slag formed an arc of black hills, with pools of lime-green water at their feet. The only evidence of nonhuman life was the bone-white birch trees thrusting upwards through the slag. Burtynsky's first thought was that he had somehow entered an alien world. The black land 'totally destabilized me', he said. 'I thought, is this Earth?' But he quickly came to realize that what he found in places like Frackville was the consequence of our addiction to fossil fuels, scoured into the rock and sediment. We are in the shadow places even when we don't realize it.

Burtynsky's images follow the appetites of the insatiable road. In 2007 he photographed open-cast mines in the West Australian gold fields. In one picture of the salt pan landscape around Lake Lefroy, a deep, belly-like crater is sunk hundreds of feet into the earth like the King of the Road's stomach in Okri's fable, starkly black against the white crusted plain, its ribbed layers picked out eerily in salt. In another, of the 'Super Pit' near Kalgoorlie, the mine's true scale becomes apparent only when you notice the small town perched on its rim like a speckle of white lichen. The pit's mouth yawns 3.5 kilometres across at its widest point, and plunges down 180 metres. A squiggle of access roads winds down to its tapering base. The mine extracts around twenty-three tonnes of gold per year, but every gram of gold involves the displacement of half a tonne of earth.

Our capacity to move sediment like this, far exceeding the rates produced by geological processes, will leave behind countless future trace fossils, both large and small, from vast craters like the Kalgoorlie Super Pit to the minerals themselves, which we have drawn from deep in the earth and scattered over its surface. Increased concentrations of gold, copper and platinum at the surface, as well as of the toxic heavy metals produced during mining such as cadmium, lead and mercury, will bear witness to the way we have relentlessly sought to feed our appetite for prized minerals. Geologists speak of our ability to distribute rocks and sediments far from their place of origin as comparable to the way glaciers drop erratic stones in distant valleys.

'When you think about it', writes Michael Mitchell about Burtynsky's images, 'there's a big hole somewhere for every stone building on the planet.' Burtynsky's photographs of the Rock of Ages quarry in Vermont resemble the urban canyons of a city like New York. Cut into regular horizontal sections like the floors of a skyscraper, the ledges highlighted by a fortuitous snowfall, it seems as if a completed building has been lifted from the earth. Meltwater has streaked the grey granite walls to obsidian black. Michelangelo famously insisted that the statue is already inside the stone and that it is the sculptor's task to reveal it (the first quarry Burtynsky photographed was Carrara in Italy, where the marble for Michelangelo's statue of David was extracted). In Burtynsky's photos, we are confronted with a kind of ghost architecture, as if our cities were somehow excavated whole.

IN 1997 BURTYNSKY HAD what he calls an 'oil epiphany': that all the many manufactured landscapes he had photographed 'had been made possible by the discovery of oil'. He resolved to trace with his camera the immensely complex infrastructure of fossil fuels, from extraction to the depleted fields and devastated landscapes it leaves behind. A few weeks after my trip over the old Forth Road Bridge, I visited the university library to look at *Oil*, the book that gathers together the results of Burtynsky's quest.

Much of its first half has a heroic tenor. Wide-angle pictures of California oil fields are filled with a boundless herd of thousands of nodding derricks, docile and bovine, cropping the desert as far as the horizon, recalling the lost buffalo herds of the American plains. In Burtynsky's images of oil refineries, miles of glistening pipes are crammed into the frame, arterial skeins that hint at symmetry but also elude it. Some show the world made possible by oil. In *Highway #5*, an aerial shot of the intersection between Highways 105 and 110 in Los Angeles (where the opening song-and-dance number to *La La Land* was filmed), the road is elevated to mythic proportions. The city's sprawl fills the picture right to the edges, and as far as the San Gabriel Mountains to the north. But everything is dwarfed by the road – the endless rows of paperlike houses, even the huddle of downtown skyscrapers, are diminished, dominated by the prodigious, oesophageal multi-lane highway unfurling from the intersection and bulging and twisting away northwards.

The image reminded me of the work of the science fiction writer J. G. Ballard, who claimed to prefer concrete landscapes to meadows and predicted that the freeway system will be all that remains of Los Angeles once the city has faded into memory. People of the future, Ballard suggested, will see its slipways and overpasses as enigmatic testaments to our standard of beauty, just as we look admiringly on the mausoleums of Giza.

Other pictures ask us to imagine oil's shadow worlds. Burtynsky turns his camera on an exhausted oil field in Baku, Azerbaijan, the site of perhaps the world's first industrialized oil field and the largest in the world at the start of the twentieth century, from which oil had been dug since at least the fifteenth century. Skeletal derricks and towers, emaciated and blackened or rust-blushed, take the place of the placid, nodding machinery in the Californian photos. In the foreground,

angular metal debris points to the sky like the ribs of desiccated cattle.

Ten million years from now, every human structure that
exists on the surface of the earth will have been worn away.
The largest and most extensive trace fossils of us will be underground. Animals can burrow to a depth of around two and
a half metres; the deepest plant roots reach less than seventy
metres. Humans, by contrast, have dug holes deeper than any
life-form has ever gone before: the Kola Superdeep Borehole in
Russia's remote north-west is only twenty-three centimetres in
diameter, but it descends more than twelve kilometres into the
ground, far below even the deep-dwelling microbial communities that inhabit pockets and fractures in the rock up to five
kilometres down. And we have done so everywhere. Beneath
each grazing Californian derrick and pinched, leaning tower in
Baku is a borehole perhaps up to a kilometre deep; worldwide,
there are thousands of similar boreholes on every continent
save Antarctica. The Anthropocene Working Group estimates
that, laid end to end, they would make a borehole around fifty
million kilometres in depth – equivalent to the total paved road
network, or seven metres for every person alive today. Whereas
surface roads will be preserved only in fragments, these boreholes will be sheltered from erosion. Some may be crimped and
compressed by metamorphic processes, or gradually raised to
the surface and weathered to dust, but others will be there in
perpetuity, columns opening deep into the earth, coated in
residual oil and barium-laced mud. Closed mines will lead to
huge subterranean voids where, in our drive for coal, we have
removed entire strata.

The insatiable road will be the source of some of the most

telling future fossils. Even in fragments, they will hint at the extent of our reach across continents and into every wilderness. An astute observer of these splinters might be able to pull together an even larger story, of thriving megacities and planet-spanning industries, of our thirst for fossil fuels and the depths to which we have pursued them. Even more incredibly, beneath the sea some of the longest roads, passing between the continents, may survive intact.

In the final section of *Oil*, Burtynsky visits ship-breaking yards in Chittagong on the coast of Bangladesh. A series of dangerous accidents in the late 1990s and early 2000s involving heavy-fuel oil tankers off the French coast led to the banning of single-hulled tankers and spawned a new industry, as dozens of ships were beached at Chittagong to be broken down for scrap and recycled. In Burtynsky's pictures, the partially dismantled tankers have assumed sculptural or even topographical forms. Ships minus their prows reveal their hulls in cross-section-like strata in an iron cliff; some have become metal escarpments and overhangs. It is a desperately precarious industry, undertaken by barefoot men using little more, Burtynsky says, than 'torches and gravity'. Injuries and fatalities are commonplace, and their working environment is a malign concoction of oil and flakes of highly toxic marine paint. As the ships are broken down, they can yield as much as fifty thousand metres of copper cable, dozens of kilograms of aluminium and zinc, and tens of thousands of litres of oil. In the process, the ships are whittled away to almost nothing.

And yet, even before they come to rest on the Chittagong beach, they have played a part in creating future fossils. Although dumping shipping waste in the ocean was banned in 1972, it's estimated that more than six hundred thousand tonnes enters the water every year, mostly hard and soft plastic, tin cans, and fishing gear. Some will be carried by currents and end up in debris traps in submarine canyons or seafloor depressions, but enough will likely remain concentrated around the routes of major shipping lanes to indicate that this was once a watery road. These plastic seafloor trails are laid on top of layers of hard clinker, the coal-burning residue dumped over the side of steamships in the nineteenth century. Many ships also disposed of extra clinker in port when they cleaned their boilers. These tough pavements connect the major port cities of the nineteenth century like Liverpool and New York, and have already been covered by sediment, preserving them from erosion. Unlike the road network on land, which will leave only fragmentary clues about its extent, a future geologist will be able to reconstruct much of the major shipping network from these clinker roads beneath the sea.

I peered at Burtynsky's pictures of oil and its afterlives for two hours, scribbling notes and impressions, until hunger finally broke my concentration. But as I turned the last pages, making ready to leave the library, the final image struck me totally still. The camera peers directly down from ground level at a shuffle of footprints pressed into the Chittagong mud just as the Happisburgh footprints had been, eight thousand miles away and over eight hundred thousand years earlier. But these ones gleamed blackly. The mud had dried and split, and oil spilled from the broken tankers had flowed through the cracks to fill the footprints just exactly to the brim.

TWO

———

THIN CITIES

Although I was climbing up the stairs, I had the odd sensation that I was sinking underwater.

It was late May, and the trees around the university were loaded with cherry blossoms. The students had recently finished their exams and my days were a little less full, so I had decided to take a look around the university gallery. The Talbot Rice Gallery is not easy to find, tucked away in a corner of the Old College and reached through an anonymous grey door that leads to a long staircase. This is the oldest part of the university, and its stone foundations bear the tidemarks of centuries of Edinburgh weather. As I climbed towards the gallery, on the fourth floor, the whitewashed space filled with the sound of moving water; a submarine gargle, heavy and blunt, broadcast from speakers mounted discreetly in the corners of the stairwell. The higher I got, the deeper I felt I was descending.

Inside the gallery, I made for a doorway in the opposite wall and stepped into the unlit space beyond. The room was entirely dark, with a single bench in the centre set before a large screen. The video was on a loop and had already begun when I took my seat.

On the screen a spotlight was tracking low along an ordinary American suburban street. The sky was a dense, pitiless black, and the houses shone as in a photographic negative. The road was broad and spacious, but strangely uneven – lumpy and furrowed, with tall, luminously white drifts, whether of sand or snow I couldn't tell. Mournful strokes of a cello played above an electronic drone, punctuated by an occasional submarine ping. The camera continued to track along the street and then suddenly executed an astonishing dive, plunging through the road as if beneath the waves, and turned like a swimmer to look up at the shimmering houses from below. The effect was breathtaking and tinged with vertigo.

The installation, by Asad Khan and Eleni-Ira Panourgia, was named after the coordinates for the city of New Orleans: *29.9511°N, 90.0715°W.* It consisted of a 3-D animation of data collected by the U.S. Geological Survey, NASA, and the US Army Corps of Engineers using light detection and ranging (lidar) technology, after Hurricane Katrina flooded the city on 29 August 2005. Lidar creates maps by scattering pulses of light and translating the length of time it takes for the light to strike an object and be reflected back into a measure of distance. It gave the streetscape a grainy texture, as if it weren't just awash with sediment dumped by the departing waters but actually composed entirely from accumulated flecks. The buildings were nearly translucent, their shapes picked out by white dots like iron filings on a magnet. Shining white mud was heaped up on the windscreens and wrapped around the wheels of cars.

Trees and telegraph poles leaned at crazy angles. Although still standing, the houses looked strangely washed-out and insubstantial. The thin walls of a church revealed rows of empty pews, but there were no people or signs of life anywhere. Lit by the powerful spotlight against the looming dark, with clouds of sediment fogging the view, the abandoned streets reminded me of footage of seabed exploration.

Hurricane Katrina was one of the most powerful storms ever to make landfall in the United States. It arrived at 6.10 a.m. and took just twenty minutes to overwhelm the city's defences. Multiple failures of the levee system meant that some areas were inundated in a matter of minutes. At the peak of the flood, around 80 per cent of the city stood under three metres of fetid water. One thousand, eight hundred and thirty-three people died during the storm and in its aftermath. More than half were African American; 60 per cent were elderly. Upwards of a million people were displaced from their homes. The total cost of the damage was more than a billion dollars.

New Orleans is built on a deep base of waterlogged clay and silt, deposited by the Mississippi River over thousands of years. Many of the levees failed because they were built on shifting sands. Barriers collapsed as the water ate away the soft soil beneath; some of the older barriers had sunk nearly a metre below sea level before the storm arrived. The weight of the city has been, quite literally, too much for the land to bear. Subsidence was first noted in the late nineteenth century. An increasing thirst for groundwater, which creates subterranean pockets that are then compressed by the land above, and upriver damming of the Mississippi, which prevents the replenishment of sediments, have undermined the city to the point that it is now thought to be subsiding by up to 12 millimetres per year. Many other major cities built on soft deltas face the same problem. Since 1900, Bangkok has sunk by 1.6 metres, Shanghai by 2.6 metres, and eastern Tokyo by an astonishing 4.4 metres. New Orleans is sinking around four times faster than the sea is rising. About half the city is already below sea level, at its lowest point by up to 2 metres.

On the screen, the video continued to play on its loop. Again and again, I watched the camera track along the shining abandoned street, then dive beneath the surface of the thin city.

THE WATERS SUBSIDED, and people returned to New Orleans. But since Katrina, Hurricane Sandy has devastated the Atlantic seaboard and Harvey has laid waste to Houston. The Intergovernmental Panel on Climate Change projects that the population exposed to hundred-year coastal floods will increase from 270 million in 2010 to 350 million by the middle of the century. The seas rose by an average of 1.7 millimetres per year between 1901 and 2010, but the majority of the rise occurred in the past several decades; from 1993, the annual rate was nearly double the average. By the end of the century, the mean global sea level could be as much as a metre higher than it is today, a statistic with dire consequences for island nations like Tuvalu and the Maldives, but also for cities like Bangkok (one metre above sea level), Singapore (at sea level) and Amsterdam (in places, two metres below sea level).

Water expands as it gets warmer. Thermal expansion

accounts for half the total sea level rise to date; much of the rest is down to melting glaciers. Other factors play their part, and mean that the sea won't rise to the same height or at the same rate everywhere. During the warm phase in the El Niño oscillation, average sea level in the Pacific can be raised temporarily by as much as 40 centimetres. As ice is lost at the poles, the gravitational pull of the ice sheets weakens, so that the ice lost in one hemisphere contributes to higher seas in the other. Melting Greenland ice will press against the seawalls of Singapore and Kowloon. While the majority – around 95 per cent – of coastlines face rising water, around the poles the sea level could actually fall, especially in the northern hemisphere, where some landmasses are still rebounding from the weight of glaciers lost at the end of the last ice age. In New York, which has passed the apex of its glacial rebound and is now sinking on its immense bed of five-hundred-million-year-old schist, water from the West Antarctic gets caught in a bottleneck where the Gulf Stream slows along the Atlantic coast of the United States, creating a bulge of water around the city.

No one knows how quickly the oceans will rise. But all the ancient stories of the flood, from Noah to Gilgamesh, emerged during the most recent interglacial period, when it is thought the water rose by between one and two metres every century.

In *The Epic of Gilgamesh*, even the gods fear the rising waters, fleeing to the highest heavens to escape. The human cost of extreme sea level rise is terrifying. Around 10 per cent of the world's population lives in what is known as the low-elevation coastal zone, on land less than ten metres above sea level. Something in the region of 600 million people live in coastal areas with limited or little resilience to flooding. Without any adaptation, between 72 million and 187 million people will be displaced by sea level rise by 2100. Perhaps we will respond soon enough to protect threatened lives and livelihoods. But the sea is patient. The quantity of carbon in the atmosphere already means that further warming will occur for hundreds of years to come. Some scenarios predict between one and three metres of sea level rise *per degree of warming* above pre-industrial levels.

Sea level rise of this kind would necessarily involve the collapse of one or both of the Greenland and West Antarctic ice sheets. One is being eroded from above, the other eaten away from below. As warm air, coupled with soot from industry and forest fires, darkens the surface of the Greenland ice and hastens its melting, in West Antarctica warmer water is steadily eroding the grounding line binding the immense ice sheet to the continent. Beneath Thwaites Glacier, a cavity 350 metres deep and more than half the area of Manhattan has opened up, equivalent to fourteen billion tonnes of ice. These billions have not, fortunately, added to global sea level, as the ice was already underwater when it melted. But Thwaites itself is 120 kilometres wide: alone, it could raise global sea level by 0.6 metres, and it also acts as a brace on the glaciers surrounding it, the loss of which could mean that the oceans rise by a further 2.4 metres.

In total, enough ice is locked away in the planet's ice sheets and glaciers to raise the seas 60 metres higher than they are today. If all this ice were to melt, the oceans would redraw the world map. North America would shrink westward; South America would be consumed by the growth of vast inland deltas. The UK would be an emaciated version of what we know today. In Australia, water would flood through the Spencer Gulf toward the continent's red centre. China's coastline would retreat as far as Beijing, 150 kilometres from its current position.

The complete loss of all ice is a fantasy, but the collapse of the Greenland and West Antarctic ice sheets is not, and together they would raise global sea level by eleven metres. Sea level rise on this scale would take centuries, even tens of thousands of years, but if we reach the point where collapse is irreversible, it would seal the fate of every coastal city. Their litany would be like a lament for world culture. Kolkata. Jakarta. Shanghai. London. Copenhagen. Algiers. Lagos. Miami. New York. Houston. New Orleans. Cancún. Buenos Aires.

ALL CITIES ARE INCIPIENT RUINS. The ruin is there already, beneath the shining street. This is nowhere more evident than in the philosopher Walter Benjamin's Arcades Project, left incomplete on his death in 1940. Benjamin worked on his meditation on the iron-and-glass-covered arcades of nineteenth-century Paris for thirteen years (rumours persist that the final manuscript disappeared from Benjamin's briefcase when he died fleeing Nazi-occupied France), and, despite consisting of no more than a compost of notes and sketched reflections, it has been read by many as *the* essential work on the modern city. What has come down to us is itself a kind of ruin: an odd collage of memory and learning, quotation and anecdote. Benjamin's vision of the city takes shape in a great smash of detail.

The inspiration for Benjamin's reflections is Baudelaire's

flâneur, the wandering man of the crowd whose gaze pierces the city's surface and pieces its fragments together. Benjamin's Paris is, he declares, the 'sunken city' of Baudelaire's poems, with its 'sea of houses' like 'multistorey waves', 'more submarine than subterranean'. But what most captured his imagination lay submerged beneath the arcades' glass roofs. In an early passage, written in 1928 or 1929, Benjamin recalls his childhood enthusiasm for reading encyclopedias, particularly the colour illustrations of prehistoric landscapes – wild Carboniferous jungles, or 'Lakes and Glaciers of the First Ice Age'. A similar panorama unfolds, he suggests, when we contemplate the Parisian arcades, which for Benjamin resembled the sites of relics out of deep time. The consumer haunting the cavernous arcade is 'the last dinosaur of Europe'; and 'on the walls of these caverns, their immemorial flora, the commodity' persists in 'the most irregular combinations'. In the abundance of the arcades, 'a world of secret affinities' emerges, writes Benjamin: 'palm tree and feather duster, hair dryer and Venus de Milo, prosthesis and letter-writing manual', connected imperceptibly by the lives they passed through.

Benjamin's fundamental insight in the Arcades Project is that every city is composed of countless secret affinities. Materials from across the globe pour into cities in the concrete, brick, and steel of their buildings, or as coffee cups and credit cards, fibre-optic cables and window glass, diamond rings and paper clips. However separate they may appear from one another, they have a common connection. Their secret is us, whose lives they populated and intimacies they shared. More than any other trace we will leave behind – the grey roads spread over the land like veins, the deep pits we have excavated and the chemical residues we've left in the air, the ice, and the water, or the long-lived plastics and even-longer-lived radionuclides – our cities will form the most concentrated and revealing archive of who we were and how we lived. Our 'immemorial flora' will persist in the fossilized remains of our buildings, in their buried infrastructure, and in countless discarded small objects, like a vast encyclopedia of human lives and desires.

More than half the world now lives in cities; until 1800, it was 3 per cent. There were 512 cities with a population of more than one million people in 2016. The United Nations predicts that in 2030 there will be 662, adding a further seventy-two million people to the global urban population every year. One hundred and forty-five million people live on coastlines that are less than a metre above sea level. The majority are in megacities like Jakarta, Lagos, New York, and Mumbai. The number of such megacities, with populations of ten million or more, doubled between 1995 and 2015, and each one is growing. By 2030, the population of Shanghai will have increased from twenty-four million to over thirty million; Mumbai's population will have risen to twenty-seven million. Nine million extra people will live in Dhaka; eleven million more in Lagos. In recent decades cities have spread across deserts, like the shining palaces of Dubai, and colonized thousands of acres of land wagered from the sea; in the future they may burrow underground, like the multilayered subterranean city planned by Singapore.

For a limited time, all cities will leave a trace. Not a complete picture: lacking much durable infrastructure, the million or so people improvising city life in Dharavi won't contribute as much to the imprint of Mumbai as those who dwell in the skyscrapers of Nariman Point. But with large structures and deep foundations, even empty cities will endure for thousands of years as islands of concrete and glass, connected by a network of tributaries (railways, roads, sewers and pipelines).

But if we consider the fate of cities on a multimillion-year timescale, those located on higher elevations or where the ground is rising will eventually be worn away to nothing. Cities that are preserved will have been protected from erosion by the benediction of water and the balm of mud. Low-lying megacities sited on coastal plains, marine estuaries, or river floodplains, vulnerable to rising seas, have the best chance of being fossilized. Once drowned, the abandoned city will be caulked by a thick layer of mud that shields it from the eager appetites of weather and oxidation. Eventually its buildings will collapse, but what is buried, the subterranean traces – the concrete rafts and piles that hold the skyscrapers of New Orleans in place or even the submerged stone-capped timbers beneath Venice; the metro lines, the pipes and the cables – will come to form what the geologist Jan Zalasiewicz, the chair of the Anthropocene Working Group, calls the 'urban stratum', a rich layer of human traces and secret affinities compressed in the rock. After one hundred million years, what remains of New York or Mumbai may be a deposit no thicker than the shallow end of a swimming pool. Ironically, the water that forces the abandonment of coastal cities will also ensure their future.

In the meantime, the loss of the cities we know will give rise to others as people leave in search of dry land: new Miami, new Dar es Salaam, new New York. As the waves break over the old cities, these new cities will flood towards higher ground, driving their foundations into the strata and building their own worlds of secret affinities.

OF COURSE, before they are abandoned, many cities will try to appease or resist the rising waters.

Venice has been married to the sea for a thousand years. Every year on the Feast of the Ascension, a procession of boats rows out into the lagoon led by the doge, the chief magistrate of the republic, and the patriarch. At the mouth of the lagoon, the patriarch breaks an ampule of holy water over the waves, while the doge removes a gold ring from his finger and drops it overboard, proclaiming, 'We espouse thee, O sea, as a sign of true and perpetual dominion.' Their intention is to placate the waters and maintain the equilibrium on which their prosperity depends.

Identified in Renaissance iconography with Venus, the sea-born goddess, Venice is a city built on water. In the sixth century the Roman historian Cassiodorus described the Veneti, the original Venetians who established their island dwellings having fled the collapse of the Roman Empire, as 'like aquatic birds, now on sea, now on land'. Peter Ackroyd remarks that the sea flows through the city's fabric, not only in the famous canals and waterways but also in the gently undulating floor of St. Mark's Basilica and in the glassware, 'sea made solid', through which Venetian fame spread for centuries. Perhaps no one gave more thought to Venice's association with the sea than John Ruskin, whose three-volume *Stones of Venice* was published in the early 1850s. 'The Venetian . . .

built his houses, even the meanest, as if he had been a shell-fish', wrote Ruskin, for whom the city was an immense sea-shell in reverse, made 'roughly inside' but 'mother-of-pearl on the surface', the smooth classical facades 'glistening like sea waves'. 'You might fancy early Venice one wilderness of brick', he speculated, 'which a petrifying sea had beaten upon till it coated it with marble: at first a dark city – washed white by the sea foam.'

Many writers have perpetuated the myth that Venice was founded by divine decree. But for Ruskin, the city was born through a more prosaic, providentially narrow aperture: eighteen inches, to be precise, or forty-five centimetres, the average rise and fall of the tides in the Venetian lagoon. Deeper currents, he remarks, would have kept the islands apart, exposing them to invasion; stronger surges would have required Venetians to exchange the refinement of their architecture for 'the walls and bulwarks of an ordinary sea-port'. Cancelling the tide would have left the city's waste to stagnate in the narrow canals; and just forty-five more centimetres between flood and ebb 'would have rendered the doorsteps of every palace, at low water, a treacherous mass of weeds and limpets'. If so, then Venice's contract with the tides is unravelling. The tidal dynamics of the lagoon have been altered by dredging and land reclamation, and as in New Orleans, groundwater pumping has caused parts of the city to subside. At its highest points Venice is less than one metre above sea level, but in the last eighty years there have been seventeen floods of a metre or more. The greatest, the cataclysmic *acqua alta* of 1966, was nearly two metres. The city has been accustomed to flooding since its beginning, but what was a delicate balance has become dangerously weighted. Half the major floods have happened since 2000. Whereas once Venice might have encountered limited flooding less than once a month, now part of the city is underwater seventy-five times a year.

Since the mid-1970s, galvanized by the shock of 1966, Venice has been intent on constructing a defence against the rising water. The MOSE barrier (Modulo Sperimentale Elettromeccanico, or Experimental Electromechanical Module) is a set of submerged inflatable gates fixed to the bottom of the lagoon that can be raised during high tides to separate the lagoon from the sea. But whereas its near namesake parted the waves of the Red Sea, the MOSE has been designed to cope with no more than a twenty-centimetre increase in sea level. If the IPCC predictions for global sea level rise come to pass, then the MOSE barrier will have failed even before it is opened. The seas could overtake it by 2050.

In Italo Calvino's novel *Invisible Cities*, the Venetian explorer Marco Polo is granted an audience with Kublai Khan, who presides over a colossal, ruinous empire. The khan invites Polo, whose travels have carried him across the length and breadth of the empire, to describe the cities he has visited. Cities, Polo informs him, are like dreams. Both can take whatever form can be imagined, but even the most unexpected dream conceals submerged fears and desires. The khan's desire, and fear, is to see his empire whole, to know its extent. So Polo provides him with a fabulous catalogue of impossible cities – cities whose museums contain models of every possible alternative version of themselves, or whose streets are laid out like a musical score – and yet it becomes apparent to the khan that, despite the astonishing breadth and variety of his vision, Polo can speak, endlessly, of only one city. 'Every time I describe a city I am saying something about Venice', he confesses.

Among the clusters of different kinds of cities he describes are what Polo calls 'thin cities'. Cities like Isuara, whose border repeats the outline of a deep subterranean lake, or Zenobia, a city raised on stilts. Some are remarkably similar to the various strategies, like the MOSE barrier, which engineers are devising to defend their cities against the rising seas. New York has proposed to build the 'Big U', a huge seawall around lower Manhattan that would protect the financial district but leave anyone who lives north of West Fifty-Seventh Street exposed to the waves. Along the coast of the Netherlands and in the Mississippi Delta, engineers have constructed artificial islands that will deliver regular quantities of new sediment to the eroding coastlines. To protect Rotterdam, which is already 2 metres below sea level, engineers have combined massive barrier systems, like the 210-metre-long steel gates of the Maeslantkering at the mouth of the Rhine, with designs for floating houses that will rise along with the water.

Many of the cities Polo describes are haunted by their alternatives. There is Clarice, the palimpsest city, built and rebuilt time and again by recycling the fabric of the old city, and Laudomia, a city accompanied by its past and future, in whose infinite architecture of niches and fissures unborn generations crowd every possible space. Cities threatened by rising seas, similarly, are ghosted by their possible future selves, drowned and relinquished or resilient and thriving. I can only suppose that to live in one of these cities would be to feel already displaced, as if the city of the future had already abandoned me.

Like Polo's cities, Venice has its shadow. Ruskin rhapsodized about the incorruptible beauty of Venice, 'which seemed to have fixed for its throne the sands of the hour-glass as well as of the sea'. Yet he also saw Venice as 'a ghost upon the sands of the sea', so lost in its decline that 'we might as well doubt, as we watched her faint reflection in the mirage of the lagoon, which was the City, and which the Shadow'. Venice, a city founded by refugees caught between the tides of eternity and entropy, may be one of the first to be abandoned to the waves. Yet when we speak of Venice, we do so in the shadow of not only its own watery future but also those of Dhaka and island nations like Kiribati. And perhaps in the future, like Marco Polo, whenever we describe a city lost to the sea, we will also know that we are saying something about Venice.

JINLING ROAD WAS FILLED with music. The street is lined with shops selling musical instruments, and as I walked past each door the warm tones of violins, the liquid sounds of teardrop-shaped Chinese pipa, or the corvid blast of an electric guitar broke over me.

The music of the instrument sellers was a relief from the bright horns and coughing motors of dozens of mopeds weaving between the traffic, their riders often talking animatedly on mobile phones, even mounting the pavement and leaning on their horns to clear a path. Green lights seemed less like an indication of a right of way than a signal to pedestrians and motorists to open negotiations. W. G. Sebald contrasts the experience of waking to the stillness of Venice with the tumult of other cities. The sound of rushing traffic is, he says, 'the new ocean', breaking in surges 'across the stones and the asphalt'.

Shanghai's streets are washed by waves of traffic noise, but the real music of the city comes from construction. The rapid pulse of drills, the whine of saws and the clang of hammers on steel seemed to accompany me wherever I went.

Since the 1840s, when the city was little more than a thin strip clinging to the muddy western bank of the winding Huangpu River, Shanghai has expanded its total area to more than six thousand square kilometres. The population was twenty-three million people in 2010; the number of sky-scrapers exceeded eight hundred. Both are increasing exponentially. Situated on bluffs above the Huangpu, a tributary of the Yangtze, Shanghai – which means 'above the sea' – is sinking. Uncontrolled extraction of groundwater has hollowed out the terrain beneath it, leaving the city to subside at an alarming rate: a total of 2.6 metres since the problem was first noticed, in 1921. Remedial efforts to protect the city include periodically raising its concrete floodwalls and pumping water back to 'reinflate' the ground (which has elevated some areas by eleven centimetres – although in 2012 a Chinese geological survey reported that replenishing the one hundred billion cubic metres of water withdrawn from the aquifer since the 1970s would take around ten thousand years). But its cosmic-looking skyscrapers are built on concrete-and-steel foundations pushed up to ninety metres into the mud, and at over five hundred kilometres, its metro system is the longest in the world.

Rapid expansion and deep foundations mean that the city has already left its mark on the strata. I had come to Shanghai to see for myself what is guaranteed, in time, to become an immense future fossil.

The Scottish poet Hugh MacDiarmid called Edinburgh, draped over the black rocks of ancient volcanoes, 'a mad god's dream'. According to Daniel Brook, Shanghai was also the product of a mad dream: to build a global trading city in the midst of one of the most self-sufficient empires in the world. In 1793 the Qianlong emperor dismissed British overtures to establish trading relations as unnecessary. In China, he explained, 'we possess all things'. But the treaty of Nanjing opened five Chinese coastal cities to foreign trade in 1842 and set in motion a sequence of events that would make Shanghai one of the world's greatest cities.

Few cities have reached megacity status so quickly. Shanghai has experienced a series of convulsive expansions, sometimes in abeyance, sometimes in flood. The first was a land boom that lasted from the 1840s to the 1860s, when hundreds of thousands of refugees from the Taiping rebellion against the Qing dynasty poured into the city, making it the fastest-growing on the planet. A further surge meant that the population doubled between 1895 and 1915, and by the 1930s, Brook writes, Shanghai was 'the most modern city in the world', a Jazz Age riff on art deco style, architectural ambition, brutal gangsterism, and rapacious consumption. With a population of more than three million people, it was the sixth-largest city in the world by 1934, and also one of the most densely populated, with six hundred people per acre of space. In his memoir *Miracles of Life*, J. G. Ballard, who grew up in Shanghai and was interned outside the city during World War II, recalls riding his bike along Nanjing Road past 'dragon ladies' in ankle-length mink coats and people left to starve in the gutters. In the 1920s and '30s the International Settlement was completely rebuilt as a facsimile of global cities like New York. A row of skyscrapers shot up, planted in ground that up to this point had been deemed too soft to support tall buildings.

Like New Orleans, Shanghai is built on former swampland; like Venice, it overcame the challenges of its situation by mounting its buildings on timbers driven into the pulpy ground. The city stands on a three-hundred-metre-thick wedge of unconsolidated mud and sand deposited by the Yangtze over the past three thousand years. In the 1930s, however, a new confidence overtook doubts about the terrain. An aphorism of the period crowed that 'the neurotic thinks that in fifty years Shanghai will sink beneath the horizon under the weight of these big, tall foreign buildings'. The modernist writer Mu Shiying declared Shanghai of the 1930s to be 'a heaven built on hell'. To achieve the bullish skyline of the Bund, the former colonial district, a landscape on the far side of the world was transformed. Thousands of Douglas fir trees, some of which grow to up to one hundred metres tall, were logged in Oregon and shipped across the Pacific to provide foundations for the new buildings sprouting from the doughy ground along the Huangpu.

The Japanese invasion of Shanghai in 1937, which led to Ballard's internment, put paid to this phase in the city's skyward expansion. Nationalists assumed control of the city after World War II, and during decades of Maoist rule its horizon didn't change much, although the population continued to grow exponentially. But during the 1980s an ambitious new plan for Shanghai gestated – to convert the neglected

Pudong area, directly opposite the Bund on the east bank of the Huangpu, into a glittering symbol of Chinese ambition. What arose from the marshy soil of Pudong was a skyline that could have come from science fiction. Whereas once Pudong was home to thousands of Chinese migrant workers living in improvised shacks, new Pudong would come to house more skyscrapers than Manhattan, including some of the world's tallest buildings. It was, Brook affirms, 'civil engineering on a pharaonic scale'. In Qiu Xiaolong's *A Case of Two Cities*, the poet-detective Inspector Chen Cao observes that the gap between rich and poor in new Pudong is like that 'between cloud and clod'. A million families were displaced to make way for this reimagined version of the city.

I was on my way to catch a ferry across the Huangpu, to visit Pudong and witness the sci-fi skyline. As I made my way down Jinling Road, between the snatches of music an ugly greenish smell occasionally hit my nose – like sewers fermenting in summer heat, as if to recall the swampy ground beneath the paving slabs.

The stately Bund and the futuristic Pudong face each other across a deep bend in the river. The buildings of the Bund have a muscular, imperial bearing – broad-shouldered and soberly elegant, built from impeccably dressed stone. Pudong, by contrast, is a riot. Ballard called Shanghai a 'self-generating fantasy', and when one is faced with the oneiric prospect of Pudong, it is as if the fantastic has finally melded inextricably with the real. The cluster of tall buildings glittered in the sun with a superabundant confidence. There is something gloriously eccentric about Pudong. The Jin Mao Tower is a disarming mix of modernity and classicism, a hypercharged pagoda covered in reptilian scales. The Oriental Pearl TV Tower, with its petrol-hued spheres mounted on a concrete tripod like glass onions on a skewer, blends Soviet-style austerity with magic realism. The Shanghai Tower, the second-tallest building in the world at 632 metres, is so high that it carries a warp in its structure to cope with wind stress, curving elegantly like a forearm raised to the sky, twisted at the elbow. The skyscrapers gave back rippling images of one another in their glass exteriors, multiplying the city like a hall of mirrors. The Mandarin term for skyscrapers, *mótiān dàlóu*, literally means 'the magical tall buildings that reach to the sky', but to build their fantasies Pudong's engineers still had to contend with boggy reality. To create a stable base for the Shanghai Tower, engineers poured a six-metre-thick concrete raft on top of hundreds of ninety-metre concrete-and-steel pilings.

Greasy-looking barges, low in the waterline under their heavy loads of rusted containers, patrolled the muddy yellow river, which offered thin reflections of the towers, as if to hint at their subterranean selves thrust deep into the marshy ground. The ferry deposited me in Pudong, and I made my way to the Shanghai Tower ticket office. I wanted to see the city from above, to get a true sense of its scale. My neck ached looking up at the tower, but the thought of how far it plunged beneath my feet, and the weight of the building itself pressing into the soft earth, made me feel dizzier still.

To reach the viewing platform on the top of the Shanghai Tower, I had to shuffle through a security pat-down and submit my bag to X-ray examination. I even had to drink from my water bottle in front of the security staff to prove it contained nothing dangerous. The lift carrying visitors to the 360-degree walkway is purportedly the world's fastest, and as it whisked us skyward I felt the gentle press of gravity like a hand on the back of my neck.

The Shanghai Tower is keen on superlatives, and I'd approached it feeling sceptical about such grand claims, but as we emerged from the lift, the view at the top capped my cynicism. The city was endless. Tall buildings marched off in every direction. The rococo Pudong landmarks gave way to ranks of uniform apartment blocks, until they became lost in a milky haze. To the north, where the Huangpu meets the Yangtze, I could see Chongming Island, where the steel for the Queensferry Crossing was manufactured.

One of the things I love about Edinburgh is that you can climb a hill outside it and get the whole city in view. You can see its limits, and the natural obstacles – water and mountain – that check its ambitions. But in its apparent endlessness Shanghai seemed to challenge any notion of limits. Any gaps I could see were in the process of being filled by new buildings. The essence of *Invisible Cities*, Calvino wrote elsewhere, is that in fact 'cities are turning into one single city, a single endless city where the differences which once characterized each one of them are disappearing'. Already, Shanghai is the centre of one of three super-regions in China, clusters of cities connected by high-speed rail lines that collectively are home to tens, even hundreds of millions of people. From the observation deck of the Shanghai Tower, it seemed to go on for ever. It was easy to imagine the world as one perpetual city.

A commotion made me turn away from the view. A group of children was delighting in the thrill of an LCD floor display, which gave the illusion that the floor was cracking and crumbling and opening up to reveal a plunge to the ground more than six hundred metres below us. Despite its apparent confidence, the tower seemed to be constantly engaged in imagining its own dissolution. Facts and figures regarding the scale of its construction were everywhere, apparently brandished with pride, but there seemed to be an anxiety behind this, as if they were a mantra to counteract the forces that might otherwise drag the building down. From above, the surrounding apartment complexes formed distinctive shapes almost like Chinese characters; farther out, they resolved into a Morse code of parallel linear blocks. The empty barges on the Huangpu looked like lost shoes floating down the river. I was struck by how flat it all was, stretching away eastward to the coast or spreading out west into the haze.

As the afternoon wore on, I walked round and round the circular observatory, transfixed by the view, until eventually I lost all sense of orientation in the endless whirl of the continuous city.

SHANGHAI SHAPED the imagination of J. G. Ballard. Having grown up in a place that seemed conjured from illusion, Ballard claimed that the challenge was 'to find the real in all this make-believe'. Many years after he left the city, he recalled life in the Lunghua internment camp, a few miles outside the city. He recalled peering through the barbed-wire fence across empty rice fields to where the abandoned apartment buildings of the French Concession stood 'surrounded by unbroken areas of water in the sun'. Decades later, in the early 1960s, the flooded fields and the buildings that seemed to stand in water came back to him. *The Drowned World*, the novel that made Ballard's name, is a vision of a superheated planet in which solar flares have melted the ice caps and humanity has retreated to the only remaining habitable regions, within the Arctic Circle. London, where the novel is set, is submerged beneath a series of lagoons thick with Triassic vegetation. 'I'm sure now that was the landscape I used in *The Drowned World*', Ballard said of Lunghua years later, 'though I thought I'd invented it when I was writing the book.'

The evening after my trip to the Shanghai Tower I went in search of Lunghua. I'd read that not much was left – Ballard himself returned to the city in the early 1990s and found that little remained of the camp or the districts he recalled – but the location was well known, beneath the Shanghai High School. Eighty years ago the site was well beyond the city limits, but it had long since been absorbed.

It was rush hour when I ventured out. Millions of people were on the move, and the familiar symphony of moped horns greeted me as I emerged from the metro station and set off along Shilong Road. The day was browning at the edges, and the horns were more insistent as people made their way home. Weary commuters sagged at bus stops or gathered in cafés to smoke and talk. Tomorrow's clean white shirts hung in grimy windows. As always, there were vacant lots populated by cranes and heavy machinery. After fifteen minutes or so of walking I arrived where the camp once stood, at the junction between two busy roads. When Ballard was here, the view gave out unimpeded across rice fields to the elegant apartments of the French Concession to the north. But standing on this crowded intersection I felt like I'd been swallowed by the sprawl. The horizon that had been laid open for Ballard pressed inwards, cluttered by apartment buildings. Even if I could have risen above these obstacles, I would have felt the city stretch away, mile after mile, in every direction.

Shanghai's investment in high-rise construction, Pudong's rush for height – these aspirations to inhabit the sky belie the fact that the city's real long-term preservation potential is underground. Before I had visited the observation deck of the Shanghai Tower I had eaten my lunch in the brightly lit food court at its base. Shopping precincts of various sizes can be found beneath most tall buildings in Shanghai; many metro stations also have their own malls many metres deep. This one was two levels below the surface, and there was a car park below that. Always heaving, always busy, Shanghai presents itself to the world as a city of the sky, but as I stood on the street corner, hemmed in where Ballard had been able to look beyond the barbed wire to the city's limits, I thought about the caverns beneath the tallest buildings and the endless, snaking tunnels of the metro system. So much of the city's life is lived below the streets. It is here, in the urban stratum, that the most telling traces will be preserved.

The average skyscraper is constructed from thousands of tonnes of reinforced concrete, steel, glass, plastic, copper wiring and decorative stone. The Shanghai Tower weighs 850,000 tonnes. In *The Earth After Us*, Jan Zalasiewicz details the durability of these materials. There are natural analogues for many of the main artificial ingredients for a skyscraper. Concrete, he notes, has 'inbuilt geological durability', being composed

mostly of very hard-wearing quartz, along with nearly indestructible zircons, monazites and tourmalines – some of which
will have already passed through one or more mountain-
forming cycles. Bricks are more akin to metamorphic rocks,
having been strengthened by firing; obsidian, naturally forming
glass found in volcanic rocks, tells us something of the future
of the glass that frames our cities. Other materials – steel, plastic – will exhibit more clearly the intervention of industrial processes, but over shorter geological timescales (millions rather
than tens of millions of years), they will still stand out as evidence of nonnatural processes. But it is the incredible concentration of these materials that will be the most striking aspect
of their fossilization. Coupled with the vast transport, energy
and sewerage networks that connect building to building and
conurbation to conurbation, and the municipal landfill sites
on their outskirts, our cities have the potential, according to
Zalasiewicz, to leave a trace that will still be legible one hundred million years from now.

'Burial', he promises, 'will be untidy.' Even faced with
inevitable floods, some people will be reluctant or unable to
leave their homes. Rising seas will mean rising insurance costs,
which in turn will gut the real estate market, eroding the tax
base and undermining the resilience of even the richest coastal
metropolis. The wealthy will retreat inland, leaving the poorest
to face the waters. Collapse will be gradual, as some parts of
cities are relinquished while others are saved. Some, like New
York with its enormous seawall, might survive for centuries in
this manner. But wherever the sea claims a coastline, a street
or a lone building, the story will be the same: inundation,

abandonment, and sedimentation. *The Drowned World* provides a portrait of this first stage in the fossilization of cities. Following the loss of the ice caps, Europe lies beneath several metres of water. In cities like Paris, Berlin, and London, all single-storey factories, brick-built buildings, and the sprawling suburbs are gone, and only the steel-supported towers remain standing. For hundreds of years, as the seas creep upwards, our abandoned coastal cities will look like accidental Venices, lawless and flooded places where forgotten people squat in once-elegant surroundings. Perhaps building materials will be scavenged for use elsewhere. But seawater is highly corrosive to concrete and steel. The proud skylines of Pudong and Manhattan will decay like a mouthful of neglected teeth, perhaps for as long as a thousand years, before they finally collapse.

Meanwhile, thick marine mud will wash through the submerged lower floors, protecting basements, underground shopping arcades, metro lines and everything left within them. After one thousand years the reinforced concrete pilings beneath Pudong may be up to twenty metres underwater, buried like the roots of impossible trees beneath several metres of mud and sand. Zalasiewicz describes a catalogue of curious subterranean transformations: worms and other sediment-dwelling creatures will feast on any lingering organic matter such as paper or textiles; waterlogged wood may begin the slow transformation into peat. Bricks will swell like sponges as they are penetrated by water and eventually crumble. But perhaps the most astonishing transformation will affect metals. Some, like copper and zinc, are soluble; others, like aluminium and titanium, will be protected from decay by the thin oxide coating on their surface. But the iron in reinforced concrete, in steel girders, or even present as tiny components in discarded mobile phones, in laptops, or as hair clips and razor blades, will become something wondrous, acquiring a golden colour as it reacts with sulphides in the sediment and transforms into pyrite or fool's gold.

In *Miracles of Life*, Ballard describes exploring the grounds of a derelict casino with his father in the aftermath of the Japanese invasion. 'Everywhere gold glimmered in the half-light', he writes of the now-silent casino cluttered with overturned roulette tables and fallen chandeliers, like 'a magical cavern from the *Arabian Nights* tales'. At this stage, the pressure of sediment on the subterranean chambers of our buried cities will not be enough to crush them entirely; although most will fill with sediment, some pockets might persist intact. Pyrites will form in these subsurface cavities, filling the empty space to create shining replicas of their former interiors. Any objects left within will likewise be coated with sulphides and infilled. In some cases all that will be left are fragments, but in others, entire glittering rooms filled with false treasure will be preserved.

With Zalasiewicz's account of the fossilization of cities running through my mind, I found myself wandering Shanghai's subterranean shopping malls, travelling on the metro from arcade to arcade. Most were just one level below the surface, but that's already comfortably below sea level. The larger ones plunge down several floors; all were busy, bustling places. Everything was here, from wigs to designer watches. The high-end outlets were arranged along curved, flowing boulevards, those that traded in counterfeit designer goods in functional grids, but each one glistened as if already coated in pyrite. The bright lighting and soothing music seemed designed to distract shoppers from the fact they were choosing to spend their time many metres below the surface. The marbled floors felt solid, but I couldn't shake the sense that hundreds of metres of soft clay and sand lay between my feet and the firm bedrock. I stared at the billboards and posters of beaming models advertising the latest fashions and wondered whose face would be the last to smile back at the empty arcades.

At the end of the day, as a heavy rain began to fall, I washed up at the bottom of the Super Brand Mall in Pudong. The surface was two floors above me, but standing at the base of the vaulted atrium I could see ten floors of glittering consumer opportunities towering over my head. As shoppers flowed past, in laughing pairs or fractious families, I tried to imagine the deep future of this space once it has been abandoned and silenced and the sediment begins to pour down through the atrium.

Looters will very likely take the objects of highest value, but presumably much will be missed in the rush, or regarded as valueless. Everything in the arcades seemed to leap out at me as a potential future fossil: bright pink nylon wigs and the plastic heads wearing them; fake leather handbags; make-up stations with their dozens of tiny bottles and implements; jewellery, both real and fake; miniature models of the Oriental Pearl tower; the stainless steel counters of the food court, the glazed plates and plastic cutlery; the floors of imported stone and the rows of lights; the cables, pipes and wires.

The potential for these everyday objects to become fossils

lies in their abundance. In *Invisible Cities*, Polo describes how the city of Leonia is made new every morning – its citizens wake up between fresh sheets and find their homes stocked with the most up-to-date appliances. Each evening they dispose of everything wholesale so that the city expands exponentially. Immense piles of stratified waste tower over it, and as the Leonians' ingenuity increases, the objects they devise become ever more durable. 'A fortress of indestructible leftovers surrounds Leonia', Polo reports, 'dominating it on every side, like a chain of mountains.' Likewise, throughout our lives we shed countless manufactured items, from coins to plastic cutlery, 'much', Zalasiewicz says, 'as a plant sheds pollen'. Entombed in the mud, the foundations of our cities and the teeming quantities of materials poured into them will leave distinctive traces; stainless steel, for instance, could endure long enough to leave the impression of a window frame or frying pan in the sediment. The hash mark of rebar or the curve of a hubcap will create curious shapes to be decoded; even, perhaps, entire subway trains and lengths of track will remain. But it is the potential for something as ordinary as a paper clip to become a fossil that turns the mind over. As the sediment that fills the subsurface levels hardens under pressure, the rock will be filled with the curious outlines of chopsticks, air-conditioning units, bicycle wheels, credit cards, vending machines, bottle tops, pen caps, staples, SIM cards, fake fingernails, ice-cream scoops and electrical sockets.

Even paper could fossilize in the right conditions, particularly the plasticized paper used to print magazines. What was printed on them won't survive, but there's an irony in the thought that the impression of a lifestyle or gadgetry magazine might persist, wiped cleaned of its blandishments and adverts, alongside sedimentary impressions of the objects it once persuaded us to buy. We can imagine the thrill of a future archaeologist who discovers the precise outline of a pen, a spoon, or a coil of wire in this hoard of secret affinities. But if anyone does discover these fossils, what will astonish them most is the sheer quantity.

After several million years the sediment layer will be hundreds of metres thick and weigh billions of tonnes. Most of what lies beneath it will have been squeezed and distorted beyond recognition. Some everyday fossils may be crushed but held together by the pressure; in exceptional cases whole rooms might be preserved, pockets filled with three-dimensional impressions of chairs, spectacle frames and mannequins, or doorways leading to nowhere. But for the large part what remains of our cities will be decipherable by the chemical signatures and distinctive colours printed in the sediment. Iron leached from steel will leave a bloodred stain. Water squeezed up through the rock will dissolve many of the minerals, even the calcium carbonate in cement if it is acidic enough. Fragments of artificial glass will be glazed with cataracts, like glaucomatous eyes staring blindly into the dark.

One hundred million years after the sea takes the city, it will lie several kilometres beneath the surface. Former megacities will become thin cities, reduced to a narrow stripe in the strata. A few that happen to lie in the churning engines of mountain-forming regions will be melted and twisted out of existence, but the majority will be petrified as a metre-thick layer of rubble, an urban breccia punctuated by the outlines of everyday objects, and coloured by the deep red of iron oxides, the milky sheen of devitrified glass, and the glitter of fool's gold.

AS THE TRAIN HAULED SOUTH beneath a cement-coloured sky, the mega-structures of Pudong gave way to endless ranks of uniform tower blocks – mile upon mile upon mile of identical high-rise buildings. The chain of apartment blocks was nearly unbroken, only occasionally interrupted by flooded fields or stretches of swampy-looking wasteland hemmed by small vegetable plots. There must have been thousands, each one ten or more storeys, in lockstep with marching columns of electricity towers themselves many storeys tall. Tiny clusters like opposing city-states stood proud on the horizon. Those closest to the track varied in colour from ill-looking pale greens and pinks to an overcast grey like a cigarette-smoker's pallor.

The last stop on Shanghai's metro line 16, around thirty kilometres from Pudong's crystal towers, is Dishui Lake in the new city of Nanhui, one of dozens of new cities built to accommodate China's burgeoning urban population. I was struck by the quiet as I exited the metro station. After the hustle of central Shanghai, Nanhui – originally called Lingang until it was renamed in 2012 – seemed almost deserted. A handful of Chinese tourists had travelled with me to the end of the line, but the group soon dispersed, and I was alone for the first time in days. Behind me, the broad-shouldered mass of the Chinese Customs Building squatted at the head of the sweeping main avenue. Directly opposite the station entrance was the perfectly circular, artificial Dishui Lake itself. Its eastern edge is less than a kilometre from the sea; the city itself is arranged in concentric layers that flow out to the west like ripples from a stone dropped in water.

Nanhui is a planned city, still waiting since it opened in 2003 for the overspill from Shanghai to fill its apartment buildings, and it remained a work in progress. Evidence of construction was everywhere, but around the lake the cracked paint and sea-bitten facades suggested that the elements had already begun to stake a claim to the city. Even after fifteen years it seemed largely empty, like a seaside resort perpetually out of season. A few cars drove by and I passed a small party of sanitation workers taking a cigarette break, but the only real sign of life came from a children's fun day on one of the lake's artificial islands. Kites shaped like jellyfish and squid rode the air above the festivities. Happy voices and upbeat music drifted across the still water.

Halfway around the lake, I turned off the orbital path and followed a road heading out towards the sea. The deserted recreation areas gave way to silent construction zones and then muddy fields, where tall grass grew in ankle-deep water. Once or twice I passed a solitary fisherman trying his luck at the edge of a waterlogged field. Given the absence of traffic, it seemed rather unnecessary that this road was a six-lane highway leading almost right up to the water's edge. Entropy seemed to have taken hold. Whereas the area around the lake was immaculate, here the pavement was cracked and I had to step over wrack lines of dead vegetation and broken concrete.

I was acutely aware that I was walking on land only recently reclaimed from the sea. Since the 1940s, China has appropriated huge amounts of tideland for development, adding a thick crust to the east of Pudong. With the green fields looming on either side, it felt like I was walking through parted waves.

The tide was out when I arrived at the coast, and a haze hung over the shore. The sea was just a brown smudge in the middle distance. Donghai Bridge, one of the longest bridges in the world, curled away through the mist toward Dawugui Island. I stepped down through a cleft in the seawall onto the sand. The beach was filthy, clogged with rags of plastic, used nappies and blobs of white Styrofoam like tumours. A man was gathering a pile of bamboo poles that had washed up, and nearly lost in the sea haze two figures slowly paced the tideline as if waiting for the turn. The wall was maybe two metres high, curved like a cresting wave facing defiantly out to sea. Behind it, through the cleft, I could see the ramparts of Nanhui, where the city begins its long march west.

'Until every shape has found its city', declares Marco Polo at the end of his final audience with Kublai Khan, 'new cities will continue to be born. When the forms exhaust their variety and come apart, the end of cities begins.' On the train back to Shanghai, travelling past chains of high-rises, I read again from Calvino's *Invisible Cities*: of Argia, a subterranean city where the streets are packed with earth instead of air, in which every room and staircase is like a photographic negative and of which nothing is visible at the surface; of Moriana, whose shining facades hide an obverse of rusted metal and sackcloth like figures on either side of a sheet of paper; and of Thekla, a city surrounded by scaffolding that is continually being built.

If you ask the inhabitants why its construction is taking so long, Polo informs the khan, the answer is always the same: 'So that its destruction cannot begin.'

THREE

THE BOTTLE AS HERO

One of my favourite moments in a novel doesn't really happen at all. It is an almost event, a near imagining. In William Golding's *The Inheritors*, a small party of Neanderthals living somewhere in the south of England must learn to cope with the arrival of a company of hominid strangers oddly unlike themselves. The new group is *Homo sapiens*, and they come bearing technologies that far outstrip those of 'the people', as Golding's Neanderthals call themselves. Whereas the people go naked and scavenge for food, the 'new people' arrive over the water in canoes, wear clothes and make art and music. The people inhabit a world of intuited pattern and routine. They don't yet have the mental capacity to analyse; rather, they *imagine*. Anyone in the group who conjures an idea or a new solution to a problem is said to have 'a picture'.

Early in the novel, the people have scavenged a doe killed by a large cat. They roast it by a fire they have made near the shore of the sea and begin to stew a broth in the doe's stomach. One of the party, the elder male Mal, has been injured in a fall, and as Fa, one of the young females, dips a stick into the broth to lift some of it to Mal's lips, she has a picture that leaves her, briefly, poised on the brink of another world entirely. 'I am by the sea and I have a picture', she says. A picture of the people emptying shells of seawater. Almost as soon as it arrives, the picture begins to fade, and Fa stutters out 'a meaningless jumble of shells . . . and water'. The image dissolves, and the people resume their meal; another male, Lok, runs down to the river and carries back water for Mal, cupped in his hands.

I find this moment so powerful because we see Fa come close to an idea that would, if she realized it, represent a total remaking of her world. The people live by foraging, carrying only their children and their carefully tended fire sticks. Fa's picture of using seashells to carry water away from the river, a vision that fails her before she can express it to the others, is of a world beyond subsistence, in which things of use can be stored and carried; one that can be moulded to suit the needs of the people. A more malleable, human world.

Golding's Neanderthals are affectionate and childlike, and a long way removed from the 'germ of the ogre in folklore', as H. G. Wells puts it in Golding's epigraph to the novel. Still, research done since 1955, when *The Inheritors* was published, has shown Neanderthals to be far more sophisticated than either Golding's innocents or the brute stereotype found in popular culture. Late Neanderthals possessed the skills to make stone tools and were sophisticated hunters, capable of working together to take down prey as large as mammoths. It's thought by some archaeologists that the world's oldest cave art, a precise set of crosshatched etchings on a cave wall in northern Spain, is the work of Neanderthals.

Golding overestimated the technological gulf between the Neanderthals and the new arrivals. But his novel's brief scene gestures towards a larger truth: that the long road to becoming human was marked by countless moments when our ancestors had a picture that broke through the unyielding carapace of their environment to something that could be shaped to their needs. A stone became a hammer. A shell became a vessel. Each of these innovations changed the world fundamentally. The hardest of materials could be crafted; water could be carried far away from its source. As humans have developed, we have done so in a world becoming increasingly plastic.

In 1975, an archaeologist named Elizabeth Fisher proposed that the first tool was not a stone hammer or knife, but a receptacle for moving things. Ursula K. Le Guin adopted Fisher's notion in her essay 'The Carrier Bag Theory of Fiction'. Weapons drive energy outward, she observes, but before this, there would have been a tool that brought energy home. It is fundamentally human, Le Guin says, to put something you want into a bag or basket and carry it home, to be used or shared later. The invention of the vessel broke open time and space: once our ancestors could place items they might need later in a bag, they could satisfy their hunger when and where they chose. They no longer needed to go to the river to drink or the bush to eat; they could carry the river or the forest with them. For Le Guin, this shift was the prelude to the development of culture. Our enhanced talent for subsistence, she says, provided a secure source of energy to support the massive outlay required to hunt large prey, and from these adventures, stories emerged – another kind of vessel, but for meaning.

Narratives link places and events, spinning significance out of the weave of connection; fables shape our sense of the world and of ourselves moving through it. Storytelling is fundamental to being human. Before the vessel, there was only the present and what lay at hand. But for the storyteller, the whole world is material to be scavenged and shaped into a tale.

No one can say when the first vessel was imagined, but a method for making high-density polyethylene was invented in 1953, and a patent for turning this into the plastic carrier bag secured in 1965. We know how the rest of the story goes. The US Environmental Protection Agency estimates that as many as a trillion plastic bags are used and disposed of every year, the vast majority finding their way unheeded into the oceans.

Le Guin's carrier bag theory reminded me of my last evening at a marine research station at Tjärnö, on the Swedish west coast. It was a beautiful evening, lit by the last glow of a sun-filled day, with a gentle breeze smoothing away the sharp edges of the sun's heat. The island of Saltö lay on the other side of a short bridge, and I set out, past children playing in the shallow stream and families on bikes, along the pine-scented road. Saltö is a nature reserve where people come to see deer and hare, spotted orchids and the Arctic starflower. But I had come to see what was said to be the most plastic-polluted beach in Europe.

The beach was empty when I arrived on the far side of the

island. White butterflies danced in the grass at the top of the beach, and a few gulls patrolled the shallows. At first it seemed like its polluted reputation was unearned. I saw part of a plastic fishing box and a large, white rice sack and a cracked jerrican faded to milk by the sun, but not much else. It took some time for my eyes to adjust. On the way to the beach, I'd paused to watch a pair of swallows sporting in a field by the road. Their white breasts flashed against the green, and it was only after a minute or so that I realized there were in fact dozens of birds, many more than I could count, banking and soaring over the grass. On the beach, it was the same; as I watched, what had seemed like only a couple of isolated pieces multiplied before my eyes until the entire beach was clotted. There was fishing rope, plastic cord, empty packets of Japanese trading cards, filmy wrappers bleached into anonymity and the sad remains of a child's cartoon-character helium balloon.

Most of the plastic that arrives on northern beaches is dumped by winter storms; this was August, supposedly a cleaner time of year, and the beach had been cleared in the spring. And yet it was glutted with plastic debris, most of it discarded containers of some kind. The beach faced south-west, nestled behind a hooked granite promontory that funneled plastic from the North Sea into the bay as if it was a giant bag. Peering closer, I saw that every square metre was garlanded with scraps of plastic rope and coin-size aquaculture wheels like tiny spiderwebs.

Someone had made a pile of rubbish at the top of the beach. Mixed in with the driftwood and dried seaweed were the lid of an ice-cream tub, half a dozen plastic bottles and a child's car seat. A mess of polythene strapping looked exactly like the white smash of a dead gull lying nearby. A single orange glove waved to me from the heather.

THE MORNING WAS THICKLY OVERCAST: curdled clouds and mean 40-watt light. Gulls wheeled overhead, and the fence by my bus stop was hemmed untidily with weeds and discarded plastic wrappers and bottles. I walked up to the imposing steel gate, pressed the buzzer, passed through the door ('No, to the left. The *left*', the voice through the intercom said, only slightly wearily), and received my visitor's pass in a plastic clip-on wallet. Alison Sheridan, the principal curator of early prehistory for National Museums Scotland, arrived with a warm smile and a hand outstretched to greet me.

I had come to the National Museum of Scotland's storage facility in the north of Edinburgh, not far from the shore of the Firth of Forth, where they keep items not on display in the museum in the city centre. As I had thought about plastic like the rubbish I found on Saltö, I wanted to know more about those leaps of the imagination that carried us into a more pliable relationship with our environment. Fa's picture fails her, and the novel ends with the ascent of the more sophisticated new people. But the shape of that almost-thought was realized, countless times, in efforts made by our ancestors to mould and sculpt their world, to make it conform to their will or reflect their beliefs. Having been lost for thousands of years, many of these pictured objects had washed up here, in this modest warehouse complex on the city's fringe, like the echoes of a distant rhythm. I hoped that the relics it stored might help me connect a moment like the one in Golding's novel, when a seashell nearly became a vessel, to today's oceans filled with virtually indestructible plastic.

Alison had kindly agreed to take time out of her morning to show me the collection. She had a quick, birdlike manner, with bright pink laces in her shoes and mallards printed all over her forest-green shirt. She guided me across the court-yard to another building at the back of the complex and an immense set of grey double doors, at least twelve feet high. Alison swiped a card across an electronic tab on the wall and pushed one of the doors.

The first thing I saw, standing tall and square directly in front of the entrance, was an enormous Celtic cross, darkly dignified under the cold strip lights and probably twice my height. The room behind it was rectangular and stacked from floor to ceiling with shelf after shelf of stone treasure: carved Roman heads with blunted features, grizzled brown quern stones, more Celtic crosses.

Is that cross real? I asked. 'No', Alison replied, 'it's a replica.' It turned out many of the carved stones were cast in plastic. In some cases, the original had been lost and the replica was now the only record of the artefact, and therefore, in terms of the information it provided, nearly as valuable. They were extremely fragile, Alison told me. The surfaces looked exactly like weathered stone, but ducking underneath one I saw that the base showed patches of brilliant white scarring and that the hollow interior was filled with rigid, honey-yellow foam.

Alison led me to a series of blue laminated cabinets at the far end of the room. 'Let's start with the axe heads', she said.

She pulled out one of the shallow drawers. Inside were

several dozen teardrop-shaped objects, each one resting in precisely cut hollows on a bed of charcoal-grey polystyrene foam. They were dramatically different in terms of colour, pattern, and lustre. Some were petrol grey, others peat-bleached to a ghostly ivory. One was blood red. Alison reached in and picked out a beautiful green axe, a little bigger than palm-size and deeply polished. Its glassy smooth surface gleamed enigmatically in the cold artificial light. This one wasn't made for utilitarian reasons, she explained. It had a different, much greater significance. It had been made around six thousand years ago from a green stone called jadeite, quarried from high up in the French Alps. The person who made it was seeking power, the power of the gods who lived in the mountains.

'This is green treasure from the magic mountain', she said.

She pointed out an untidy chip on one corner of the blade, the only blemish on its otherwise flawless surface. 'They did this deliberately when the object was no longer wanted', Alison said, running her finger lightly over the jagged edge. Jadeite axes were considered so potent that their makers could not just discard them, in case they fell into some other's hands. Ritually killing the axe restored the force to the holy mountain, making the object safe again.

These axes weren't tools, I realized; they were containers. They were vessels of power and protection. When that power was no longer needed, they were poured out, emptied: returned to stone.

She handed the axe to me. It was perfect, and I couldn't help thinking that not all its original force had leaked out of the chip in the blade. It had a dreamlike charisma. Alison drew out another, longer axe. This one was made of greyish tuff but with a noticeable green tone, and came from Langdale Pike in the Lake District. One edge of the axe had been left only roughly shaped, with distinct flinting scars, like a trademark. 'We think that people who migrated to Britain from the continent carried jadeite axes with them as talismans,' she said, 'carrying away some of the magic mountain to guard them in a strange land across the sea. They were broken once they had ensured a safe crossing. But people continued to value them, so they sought out other kinds of green stone.'

The treasures kept coming: drawers full of filigree-fine red-and-orange arrowheads like tiny flames, and stone balls carved with complex geometrical patterns unlike anything the makers could have seen in nature. One drawer was full of stone knives from Shetland – flat, roundish patties with razor edges, each one uniquely speckled. Any archaeological find in the UK automatically belongs to the Crown, Alison said. 'I once interrupted an auction at Sotheby's where someone was trying to buy one of the stone balls for Barbra Streisand. *But she's touched it*, they said, as if that made any difference!'

I felt astonished by how beautiful the objects were. I had expected them to be functional items, but even the tools made for everyday use showed they had been selected with an aesthetic eye. How long would they have taken to make? I asked. 'It depends', Alison replied. She estimated that an arrowhead could be knocked off by an accomplished hunter in ten or fifteen minutes, probably to be used just once, but the jadeite axe head perhaps took a thousand hours. Each object had been meticulously labelled, recording where it had been found, and sometimes by whom, the type of axe it was, and the type of rock it was made from. The jadeite axes were traded all over Europe, some travelling up to eighteen hundred kilometres from their source. Later, in the nineteenth century, collectors would trade their finds across the continent. It struck me that these were deeply social objects, that each one had a large and varied biography. However emptied of their original magic, the axes had been filled with another kind of enchantment, a rich story that took in the slow raising of the Alpine peaks and the skill of stone carvers developed over millions of years.

We moved over to a long white workbench in the centre of the room. Alison pulled a cardboard box towards us and lifted out something whitish bundled in bubble wrap.

'This is the oldest hand axe in the collection', she said, 'probably about two hundred thousand years old'. She handed it to me as carefully as if it had been made of glass. It was about the size of a flattened avocado, off-white in colour with a dark halo around the edges, and much rougher than the younger pieces we'd been looking at. The marks left from knapping the stone were still clear, each one like a ripple in water. These impact scars are called, rather delightfully, bulbs of percussion. There's a poetry in many of the terms, like *knap scatter* and *debitage*, used to describe the process of making stone tools, which is reflected in the beauty of the objects themselves. Under Alison's watchful eye, I picked up the axe and hefted it in my right hand. It felt unexpectedly unwieldy, no more than a lump of stone. But when I transferred it to my left the awkwardness immediately fell away. A gentle depression had been worn away exactly where my thumb was pressed; my fingers nestled comfortably in shallow grooves on the back. Suddenly, the object was familiar and easy; it almost asked me to strike something with it. It sat as neatly in my hand as the Bronze Age axes had fitted in their foam beds; as if the hand that once held it now held mine. The connection was startling, so much so that I quickly put it down again.

Could this have been made specifically for someone left-handed? I asked. 'It's possible', Alison replied.

Stone tools like this are the oldest extant technology on earth, having first been manufactured around two million years ago. If Fisher's first carrier bags preceded them, they must have been made from a less durable material, as none have come down to us. For most of human history, these raw implements were our sole venture into a world that we could shape and hone ourselves. Every invention that we take for granted follows from that first rock that became a tool; that leap, like Fa's, toward a more plastic world. This heritage of insights, of rough edges honed and new possibilities imagined, remains in everything we make today: the ghost in the machine, borne through their long stories. The philosopher Bruno Latour says that all made things are composites of actions and decisions that reach back into deep time. 'Most of these entities now sit in silence', he writes, 'as if they did not exist, invisible, transparent, mute, bringing to the present scene their force and their action from who knows how many millions of years past.' In handling the stone I had brushed against the margins of its immense biography. But this was true, too, of the other made objects around us – the plastic bubble wrap, the foam padding, the fake Celtic cross. Each one emerged out of a welter of past actions, and had the potential to endure far into the future. I wondered how we also write ourselves into them, and how they, in turn, write us into the world they pass through so slowly. Would someone in the far future pick up a piece of twenty-first-century plastic, something moulded to fit anonymously in the user's hand like a bottle or a toothbrush, and feel that same jolt of connection?

IN EARLY 1956, the French philosopher Roland Barthes visited an exhibition of commercial plastics. The winter was one of the coldest in memory. Blizzards confounded Scotland; Germany experienced its coldest winter in more than a hundred years. Virtually the entire Italian peninsula was smothered by up to three metres of snow, which fell, too, on the palisades of the French Riviera, and on Tunis, Algiers, and Tripoli. Ice bloomed in cracks in the Colosseum in Rome, causing masonry to flake and fall.

Since 1954 Barthes had been engaged by the magazine *Les lettres nouvelles* to produce a short essay each month on the hidden codes of everyday life. He began with the portrayal of Julius Caesar on film and went on to analyse how advertisements for soap powder, the world of professional wrestling, the brain of Albert Einstein and the face of Greta Garbo each contribute to the mythologizing of the commonplace. He was motivated, he later explained, by a sense of irritation at the way the complexities of everyday things like children's toys, or a plate of steak and chips, are disguised by the 'naturalness' of common sense. He resented the tyranny of *'what-goes-without-saying'*. These contemporary myths are a language, he insisted, but one that we most often speak without realizing it. But the greatest myth was plastic.

In the plastic exhibition, Barthes's curiosity was aroused

by a long queue of people in front of one particular trade stall, waiting patiently as if for a show or spectacle. He joined the throng, where he witnessed a cloth-capped attendant pass crystals of raw green plastic through a tubular mold to produce gleaming jade dressing-table organizers. The finished objects themselves were so banal that they disappear from Barthes's account almost as soon as they drop from the mould. What captivated him was the process, which seemed, to Barthes, wholly preternatural, a modern alchemy. Whereas medieval alchemists had once sought to derive prized metals from base sources, plastic was itself 'the very idea of its infinite transformation', a prodigy of 'quick-change artistry' equally capable of becoming a bucket or a jewel. Plastic effected a kind of transubstantiation; it was, he declared, 'the first magical substance which consents to be prosaic'.

Plastic is indeed a miraculous substance. We can bend it into the shape of our desires. Whatever the individual properties a piece of plastic might possess, however, it will almost always present them diffidently. Where the old gods demanded praise, plastic's divinity is a self-effacing presence in everyday life, so ubiquitous, in fact, that we have become accustomed to not seeing it.

As I read about Barthes's visit to the plastic exhibition, the day after holding the stone axe, I thought about how many times I had used objects made from plastic that morning. It wasn't even nine o'clock, and yet plastic seemed to have tacitly accompanied nearly everything I'd done. I had eaten food kept fresh by plastic wrappers and boxes, using implements cleaned in detergents that are stored in plastic, and which had arrived in my kitchen having passed through dozens of plastic encounters in farms, factories and in transit; I had cut bread with a plastic-handled knife, placed my kettle on its plastic base to boil water that arrived through plastic pipes and taken milk in a plastic bottle from the plastic tray in the door of my fridge to make cups of tea. I had flattened empty plastic containers and put them out to be recycled. As I do twice every day, I had popped a pill from a plastic blister pack. I had stood beneath a plastic shower head and washed using soap from a plastic bottle and brushed my teeth with a plastic tooth-brush. I had brushed my daughter's hair with a plastic hair-brush, replaced the batteries in her plastic toy and returned her clothes to a wardrobe with plastic-laminated doors. The floor I had walked on was covered in strips of laminated wood. The clothes I had put through our washing machine had released thousands of tiny plastic fibres, almost all of which eventu-ally will find their way out to sea. I had checked the time on my phone, pressed light switches and plugged plastic-covered electronic devices into plastic-covered electrical sockets. I had probably touched plastic close to a hundred times in the hour and a half since I'd woken up; by the end of the day, it could be higher than a thousand. And yet, I noticed no more than a handful of these plastic encounters at the time. Most melted into the background of my senses.

There's a photograph I show my students when I want them to think about the way plastic has influenced how we see the world. In 1986 and 1987 the photographer Keith Arnatt took a series of pictures of fly-tipped rubbish outside Miss Grace's Lane, a cave system in the Forest of Dean, in Gloucestershire. This particular photo, like the others in the series, has a fairly modest, even banal subject – a clump of weeds and wildflowers sprouting from a low platform of grey dolerite. Something is off about the picture, though – something that seems obvious in retrospect but that I remember finding disorienting when I first saw it. Arnatt had found an irregular sheet of transparent cellophane among the litter, shaped inexplicably like the continent of Africa, and placed it over the wildflowers. The cellophane lends the scene an oddly wintry aspect, as if seen through a pane of frosted glass or a sheet of ice. Rhomboids of white light are reflected in the craze of kinks and folds in its surface; yet the plastic is a curiously reticent element of the composition. Your eyes fix first on the flowers, and only then refocus on the cellophane sheet floating above them like a ghost.

Literary critics sometimes talk of the realist novel as like a window on a world, one that convinces us to forget the artifice of the novelist and accept that we are simply peering at real life as it passes by. Plastic realism has the same effect. We're so used to seeing our world through plastic that we cease to notice it.

In 1950, the amount of plastic manufactured each year was around two million tonnes; by 2015, it was four hundred million. The cumulative output for the plastic age so far exceeds six billion tonnes, and it is likely that every single piece of plastic ever produced and not incinerated still exists somewhere in some form. There are thought to be more than five trillion individual pieces of plastic in the world's oceans, much of it congealed by oceanic gyres into enormous trash islands, and it makes up a significant portion of the thousands of tonnes of space junk orbiting the planet.

We fail to notice plastic because, as Barthes saw, plastic

artefacts are entirely consumed by the present. Whereas wood and stone retain something of their origins in their textures and density, plastic is cut off from its past and absorbed by the present. Most plastics are designed to exist for us only in the moment of use. They rise when we need them and fall back when we don't. They are the subtle glue that sticks together a lifetime of detached acts. Perhaps it's for this reason they appear oddly timeless. Occasionally, a plastic object might become imprinted with a specific memory or association, like a childhood toy or a well-used tool. But the majority fade; history slides off their impermeable surfaces just as single-use plastic slips from our hand into oblivion. Plastics move in and out of our ken as if they were clouds passing across the face of the sky; their shadows cross over us without any trace of where they have come from or where they are going to.

And yet, all plastics have a story that not only reaches back into the deep recesses of the past but also pushes forward into a future only dimly perceived. For the historian of chemistry Bernadette Bensaude-Vincent, each individual piece is 'the tip of a heap of memory'. The stone tool is there in the plastic bottle, and the jadeite axe in the green dressing table organizers, but so, too, is the memory of a steaming Cretaceous world. When we let it go, we pitch it into a future that goes unregarded. If we could tell the life of a piece of plastic, what would it look like?

In 'The Carrier Bag Theory of Fiction', Le Guin describes Virginia Woolf's plan for the book that became *Three Guineas*. To write a new kind of story, Woolf realized, she needed a new language, so she drafted a revised glossary for herself: *heroism* became *botulism*; *hero* became *bottle*. 'The hero as bottle', Le Guin writes, was 'a stringent re-evaluation. I now propose the bottle as hero.'

I am by the sea and I have a picture.

THIS STORY BEGINS in mid-air. As happens many, many millions of times every day around the world, a piece of plastic has reached the end of its usable life and has just been discarded. In this case it's a bottle, made of a kind of light and durable polyester called polyethylene terephthalate (PET), but it could just as easily be a polypropylene drinking straw, a polyvinyl chloride blood bag, a sheet of polycarbonate paper, a strand of nylon fishing net or a kernel of Styrofoam packaging. One of a multitude, a single drop in the storm of single-use plastic pouring into rivers, lakes, oceans, and landfills the world over, it is entirely anonymous. Sunlight glistens on its corrugated surface as it follows a shallow arc from hand to gutter.

The bottle will take no more than a couple of seconds to fall, roughly coincident with the time it takes to be totally forgotten by whoever dropped it, and so, too, will they fall out of its story as soon as they let go. Because this is a story infinitely longer, larger, and stranger than can be told in a single moment of use. It is a vessel that contains millennia.

So we can leave the bottle momentarily, its fall suspended, while the story begins again in water covering a shallow equatorial shelf in the Neo-Tethys Ocean, in sight of the muddy coast of Gondwanaland, around 145 million years ago. Warmed by volcanism and the carbon released from the decay of immense amounts of vegetation covering the Pangaea supercontinent, our planet at the start of this story is barely recognizable. The poles are denuded of ice, and plant life thrives within the Arctic Circle. Ocean temperatures are above 50 degrees Celsius, and the anoxic waters around the equator are saturated with supercharged phytoplankton blooms. The waters of this ocean have an acid glow under the violet sky. Although shallow, this particular shelf is four thousand kilometres wide, and as far as the eye can see, in every direction, glutinous green mats form a syrupy skin on the surface of the water. The immense bloom weeps a constant stream of dead phytoplankton to the oxygenless bottom, where, free from the appetites of bacteria, it settles into the mud as a layer of preserved organic matter. For millions of years this slow cycling of algal bodies to the ocean floor goes on without interruption, countless tiny leaves drifting languorously to the seafloor. Layer is added to microscopic layer, increasing the weight that sedulously compresses the dead matter through gaps and fissures in the porous limestone, dribbling down through cracks and crevices as slowly as the leaden creep of the rock itself. Above ground, the grinding force of continental plates gradually closes the ancient ocean in which the algae once flourished, thrusting and warping the rock into accordion folds into which the former phytoplankton, now reduced to a viscid sump of hydrocarbon waxes and fats, gathers. Trapped beneath a thick cap of fine-grained sandstone, subject to immense pressure and heat, the remnants of the bloom pursue an unhurried transformation into oil.

These lightless years, the long, dark present of its formation, are beyond imagining, an imprisonment outside of time. In the world above, continents are broken apart and the oceans reconfigured. Empires of life rise and fall. Eventually – although words like *eventually* wither under the burden of timescales like these – faint tremors pass through the oil in its blind reservoir that have nothing to do with the shifting of tectonic gears. The change, when it comes, is incomprehensibly sudden in contrast to the slow pace of what has gone before. The tremors are caused by water, vast quantities drawn from the Persian Gulf and pumped into the ground to replace the pressure lost after the first profligate gushers erupted, not from the floor of the Neo-Tethys Ocean, but from the Ghawar oil field in Saudi Arabia. It floods the black chambers, reacquainting the legacy of the phytoplankton bloom with the tang of saltwater and forcing it to the surface.

Having been calmly immured beyond memory for so long, its world becomes one of flow, acceleration, and rapid change. It passes through more than a thousand kilometres of pipelines flashing under the desert sun, to a coastal refinery at Medina, where the oil is transferred to a tanker that carries it across rolling seas to a Chinese refinery.

The journey so far, from algal blooms in the glassy primitive ocean to the gleaming world of pipelines and refineries, has taken the best part of 150 million years. The next stage takes fewer than thirty days, a transformation no less radical or miraculous, when the oil becomes raw plastic and the raw plastic becomes *anything*: first, heated, cooled and distilled; then cracked into simple chemical compounds like ethylene, propylene, and butylene and recombined to form long polymer chains like rosary beads or strings of pearls. The uncooked plastic that emerges is chopped into pellets and shipped to factories that will mould it into the dream world of infinite single-use products, before we arrive, again, at the point where we began, with the same bottle falling, in a grim parody of its planktonic origins, from an anonymous hand.

As the bottle lands, it is transformed again, from a briefly useful object into waste. Perhaps it is tossed into a bin that is too full and is picked up by the wind, or simply dropped into the gutter and passes through a sewage outflow. By whatever means, its destination is assured: 8 million tonnes of plastic enter the oceans every year, much of it transported by rivers. As it happens, this bottle was discarded in Shanghai near the Yangtze River, which it enters unseen, with barely a splash. The flowing waters swiftly catch it up and carry it past other, heavier plastic objects (more recreational litter, or sourced from estuarine industries upstream) as they sink to the river bottom, where for the next decade or more they will act like cheese graters abrading lighter, more mobile plastics carried along in darkness by strong bottom currents. Suspended above a continual cloud of microplastic debris, the bottle is propelled out into the East China Sea. Before the sun has set, it is out of sight of land.

Although slightly heavier than seawater, the bottle continues to float because of a combination of surface tension and local currents and winds, which soon carry it beyond the continental shelf to the deeper, trench-riven Philippine Sea. By now, the bottle has become an ecosystem. On entering the water, it almost immediately began to acquire a biofilm of microscopic passengers in the form of bacteria and diatoms, which in turn attract a mix of creatures – hydroids, bryozoans, barnacles – that feed on the biofilm. Clouds of algae dull its transparent surface. As it reaches the deeper water, the added weight of this rich foulant layer begins to tell and the encrusted bottle sinks into the water column, where the bio-film is picked away by nibbling fish. For several weeks, as it drifts farther from land, the bottle submits to this slow rise and fall, sinking beneath the weight of each new biofilm layer and surfacing again as hungry fish lighten the load or the algae die off because of a lack of sunlight.

At last, the bottle is eased out of the knot of local currents in the Philippine Sea and picked up by the Kuroshio, or 'black stream', a fast-flowing current carrying warm tropical water northward, which shuttles it rapidly around the eastern coast of Japan. So far, bobbing and looping in the black stream's mill, the bottle has retained its shape. PET is highly hydrophobic, and although the harsh tropical sun has begun to tacitly loosen the molecular bonds in its polymer chains, the perpetual sink-and-rise caused by the accumulation and stripping of its biofilm layer has shielded the bottle from the worst effects of photodegradation and the battering chop of the waves. The strength of the eddies decreases as the bottle travels north, relaxing the Kuroshio's grip and releasing it into the eastward-tending North Pacific current, a huge conveyor belt of heat energy, nutrients, animals and plastic waste across more than eight and a half thousand kilometres of the open Pacific.

Like a pilgrim on the trail of the promised land, the bottle is now a fellow traveller in a crowd of discarded fishing nets, bottles, pipes, films, lids, bags, bottle tops and packaging materials, as well as an array of microscopic fibres, beads, and pellets. As it nears the west coast of the United States, a fraction is diverted north by the Arctic current, where it will become trapped in ice for five winters or so before finally being released into the North Atlantic. But the majority of these objects will remain under the spell of the North Pacific Current as it cycles clockwise around the twenty-million-square-kilometre expanse of the North Pacific gyre. Anticyclonic winds and ocean currents, the rotation of the planet, and friction created by spiralling layers in the water column corral the bottle and the great stream of plastics around it towards the gyre's centre. The circulating currents form an impassable boundary; the gyre is a prison it will never leave.

To many marine creatures, the coagulating soup of floating trash must look like a feast of astonishing abundance. Floating plastic bags look like jellyfish, and microplastic beads like fish eggs; Styrofoam nodules resemble algae; fishing nets coated in an epibiotic crust are plantlike. Objects at the surface are picked off by seabirds: some mistake them for food; others, like Laysan albatross, feed by scooping up water in their bills. As they graze the water, gathering up squid and fish, they also rake in small plastics that lodge in their gullets, blocking their digestive tracts so that the animals slowly starve. Beneath the surface, in the epipelagic zone where light can penetrate, gently pulsing white bags, their frayed edges undulating like tentacles, deceive hungry turtles. The majority of microplastic beads trapped in the gyre are white, clear or blue, and very similar in colour to plankton. As they sink, planktivorous fish gobble them up indiscriminately.

As the bottle turns slowly year on year inside the gyre's giant wheel, life swarms around it. Since it entered the water, it has carried encrusting organisms across incredible distances. In fact, almost all the plastic surrounding it has acted as a transport medium for species whose ranges have been steadily increased by climate change, moving organisms between continents. Drifting idly around a broad oceanic arc, the bottle collides softly with tens of thousands of pelagic plastics all colonized by hard-shelled organisms, including barnacles, coralline algae, foraminifera and bivalve molluscs. Occasionally its passage is thrown into upheaval by the wake of monolithic container ships passing through the soup, their holds stuffed with yet more fresh plastics for the world's insatiable markets and their hulls smeared with bryozoan stowaways.

The bottle has now been in the ocean for several decades. For the most part, its encounters in the gyre have been fairly benign – nudged aside by a leatherback turtle in pursuit of a plastic bag, or its underside brushed by the searching mouths of surface-feeding fish – although it has received a number of irritable pecks from unfortunate seabirds with their beaks clamped shut by six-pack yokes and threads of netting. Once it passes over a pod of humpback whales on their annual winter migration from Alaska to Hawaii. Several have tangled cords of plastic chafing angrily at their fins; each one has close to half a tonne of plastic in its guts depositing a slur of toxins in its tissues. One calf is struggling to keep up with the pod because its body has been flooded with poisons that saturate its mother's milk. Severely weakened, it has no more than a couple of days to live.

The whale calf will not make it to Hawaii, but the bottle does. One summer morning, more than three decades after it entered the water, the bottle is washed up on the southeastern tip of an isolated beach. A north-easterly wind bullies it across the black volcanic sand to where the tideline gives way to scrubby vegetation shaded by ironwood trees. The beach is crescent-shaped, a long scoop of sand bounded by a rocky headland, and very little that passes its open mouth escapes from the bowl of its belly. Cradled in this mess of cotton buds, cigarette filters, yogurt containers, burst gallon drums and hundreds of identical fellow bottles crimped into shallow V shapes, the bottle comes to rest.

In the months that follow, many of the objects will be preyed upon by starving petrels, hunger-deranged by the plastic already clogging their stomachs, their necks garlanded with gaudy debris collars. But caught in a rocky cleft away from the tideline, the bottle eludes the gaze of the famished birds. Instead, it submits to the more insistent, punishing eye of the sun. Its foulant layer soon died away once it left the water, and, washed by the rain, it has nothing to protect it from the fierce glare. The years in the ocean have already darkened it considerably, allowing its surface to absorb more heat and hastening further its oxidative degradation. UVB radiation begins to unpick its molecular knots, leaving it brittle. Milk-white cracks appear along its flanks.

At this rate, stranded on the beach, the bottle will be broken into fragments in just a few years. But its journey isn't over yet. A violent winter storm rips across the beach and tears the bottle back into the water in its chaotic backflow, pushing it around the lip of rock and once more out to sea.

As it did when it last entered the water more than thirty years ago, the bottle again begins to attract a biofilm of microorganisms. But the sun has considerably weakened it, and even though seawater shelters it somewhat from ultraviolet light, the film can no longer protect the bottle from disintegration. Waves pound it; crustaceans clinging to other hard plastics abrade it. Finally, the split seam along its side widens and the bottle breaks apart. The rough edges of its broken pieces attract more foulant, and, now considerably less buoyant, they begin to sink deeper until they fall below the point where sunlight can penetrate.

The bottle's fate is now multiple. The moulded base and neck are the first to sink. Other fragments persist near the surface long enough to be separated into smaller and smaller pieces until what remains of the bottle has become a host of thousands of microscopic particles. Because it does not biodegrade, the fragmentation of plastic can continue almost indefinitely, down to the molecular level, but most of what was once the plastic bottle finds its way to the ocean floor before then, in the bodies of dead fish or carried down by underwater currents. Larger shards and tiny particles alike circulate in undersea depressions, following the routes of submerged canyons and past seamounts decked with torn polypropylene fishing nets like vagrant kings.

There are now too many pieces of the bottle to follow individually, perhaps too many even to count, so one piece will have to represent all the others. By now, the bottle has been in the water for more than 350 years. This particular fragment, no larger than a bead on a child's bracelet, drifts across the southern face of a vast seamount below the Hawaiian ridge, which holds the Midway Islands above the waves, and into the path of a copepod – a tiny crustacean with a cyclops eye in the centre of its head and long, branching antennae. Rapidly spinning its swimming legs, the copepod generates a miniature feeding current, drawing microscopic dinoflagellates and plastic microfibres towards its mouth. The current catches up the sliver of bottle, which sticks to one of the copepod's legs along with dozens of plastic pellets and polystyrene beads gathered up by its frenzied activity.

When the copepod dies (its intestinal tract sealed by a coil of blue nylon smaller than an eyelash) its body settles in a large cleft at the seamount's base and inters the last fragment of the plastic bottle in a rich deposit of thick mud and synthetic waste. Plastic debris has collected in this cleft for the past four hundred years, encasing each fragment in an anoxic layer of sediment that will preserve it, effectively, for all time.

Burrowing creatures will mix up the layers to some extent, leaving their own trace fossil paths, but not enough to prevent a deep seam of plastic from forming in the geological strata. After this, it is a familiar, long story of heat, pressure and time. Over the coming millennia, hydrocarbons leach from the fossil plastic, accumulating in small deposits and setting in motion a slow chemical return to the bottle's origins in the rich, dark material patiently accumulated in subterranean chambers beneath the Neo-Tethys Ocean.

THERE IS A CODA to this story. As the bottle we followed began its arc toward the North Pacific gyre, an identical one was dropped into a public bin and rapidly transferred to a landfill site outside the city. As more and more discarded materials are deposited, what began as a hole in the ground grows into a small hill of waste. The oceans rise and the ground subsides, and the water begins to claim the city, but the little hill is far from the encroaching coast, and within it the other bottle endures its interment, squashed flat by the

pressure and glazed to a dull white. Protected against ultraviolet light, weathering, and abrasion, it enters a kind of long
sleep, as above ground the landscape changes from a city to
a floodplain of marshes and strange, low-lying landforms. A
river wending around the base of the hill weakens its flanks
until, one far-future day, a landslip exposes its contents, sequestered like treasure in the dim past.

With the patience of ages, wind and rain work the bottle
free until it tumbles, glinting weakly in the sunshine almost
a million years from now, into the river that will at last carry
it to the ocean.

FOUR

THE LIBRARY OF BABEL

It begins with a boy falling from a tree.

Sent to gather coconuts by his father, Tété-Michel Kpomassie is resting for a moment in the swaying upper branches. Weary from the morning's work, he lops the top off one of the coconuts to reveal the snowy flesh and drinks deeply. It is only once he has dropped the empty shell that he sees the snake, flickering with agitation by his ear. Its body is wrapped thickly around a dull cluster of eggs. Some have hatched, and the baby snakes writhe upon the mother's body.

Kpomassie begins to slide down the tree, but the snake shucks off its offspring and follows, its tongue rasping above a blazing white throat. With a wild swipe of his hand, the boy strikes out and sends the snake dropping to the ground, sliding over his hair and down his spine like a shiver. But to his horror, Kpomassie sees the animal immediately begin to climb. He is now trapped between the snake and her nest. The only option is to jump.

Although the fall did not break a single one of Kpomassie's

bones, after two days of convalescence he is still unable to stand up. His despairing father takes him to the sacred forest to be healed by the priestess of the python. As payment, his father promises the boy to the priesthood. But his fear of snakes now is absolute, and he will do anything to avoid joining the snake cult.

One morning, Kpomassie visits the Evangelical Bookshop in Lomé, the largest city in Togo. His eye is immediately drawn to one book in particular: *The Eskimos from Greenland to Alaska*, by Dr. Robert Gessain. He buys the book and takes it to the beach to read. By midday he has finished it, and his mind is gleaming with visions of a landscape of ice and eternal cold and, crucially, no snakes.

The dream of ice has captivated countless travellers, but probably none of them so suddenly and absolutely as Kpomassie. On the basis of a single morning's reading, he resolved to go to Greenland, to the very north, and to experience life on the ice among the Inuit. The journey took him eight years. He left Togo in 1958, on the eve of independence, and toiled for six years through Ghana, Senegal, Mauritania, Morocco, and Algeria. He worked his way through France, Germany, and Denmark for another two years and finally arrived in Julianehåb, now called Qaqortoq, in southern Greenland in 1965.

As Kpomassie recalled years later in *An African in Greenland*, he was an instant celebrity the moment he disembarked. All talk on the quay cut out, and heads turned towards the first African that most of the people in Julianehåb had ever seen. Wherever he went, Kpomassie was welcomed. In Europe, patrons in France financed his trip north. In Greenland, he was denied hospitality only once. He found shelter and work with ease, and whenever he left a place, he was asked when he would return. Local radio stations offered regular reports on his movements and activities. But he remained restless. Qaqortoq means 'the white one', but he was disappointed to find little in the way of ice in the south. His inner compass continued to pull northwards.

His object was Qaanaaq, also known as Thule, the northernmost town in Greenland. Kpomassie worked his way up the coast, enduring the polar winter and earning his living on fishing boats and dogsled teams. But his ambition was frustrated, finally, by what had drawn him in the first place. He waited at Qasigiannguit (Christianshåb) for a boat to Qaanaaq, but the pack ice lingered too long and no boats could make it so far north. Finally, after months of waiting, he relinquished his dream and settled for Upernavik, whose name means 'the place of spring', 550 kilometres south of Qaanaaq. His disappointment was tempered by the discovery of a traditional turfed cottage, of the kind that had been largely replaced by modern wooden houses. The owner, an old man with thick black hair snaking down his back, exclaimed that he had heard radio reports of Kpomassie's appearance in the south and had been waiting for him to arrive ever since. Robert Mattaaq agreed to let Kpomassie stay as a guest in his turf house.

Robert was an avid reader of magazines about world affairs, from which he could not bear to be parted. The piles grew larger and larger until his wife, Rebekka, demanded that they be thrown away. But Robert had a better idea. Whereas most turf houses are insulated with wooden panels that line the walls and ceiling, Robert resolved to line his with his periodicals.

Soon the whole interior was covered in a layer of articles, followed by a second, a third, and so on. Each year added a new layer, and so Kpomassie was presented with an archive of world events going back five years. Robert could dip into it to retrieve information on any given subject.

A journey that began with a fall of ten metres finally ended more than eight thousand kilometres away, on the far side of the planet. Having left Togo nearly ten years before in search of the most remote place he could find, Kpomassie arrived in the centre of the world, a library in which the past five years were compressed into a single-roomed hut on the rim of the Greenland ice sheet.

EVEN THE MOST MODEST LIBRARY conjures the image of the famous library of Alexandria, which was said to have been the most complete repository of learning ever assembled. Once knowledge was passed down through story and song, but since the invention of writing, most societies have come to put their faith in the letter rather than the litany. Libraries speak of our sense of obligation to both the past and the future: that we should preserve what remains as the surest means to care for what is to come. We assemble them because memory is unreliable, but stored knowledge is also fragile and so we value it all the more. Alexandria's library is so fixed in our imagination because, we are told, it burned to nothing two thousand years ago.

In 1901 the German mathematician and science fiction pioneer Kurd Lasswitz imagined a library without limits. By reducing all written language to the most essential elements of the Latin alphabet (twenty-two letters, the full stop and comma, and the space), he speculated that it would be possible to assemble a 'Universal Library', containing all the works ever written and to be written, or even that could be written, including all forms of error and deviation. A book may contain just a single line, or even a single letter buried deep in hundreds of blank white pages, but nonetheless Lasswitz's library would hold every possible variation: for example, every version of a book that contains just a single 'a', in each iteration placed at a different point on one page; the same again for 'aa', and 'aaa', and so on; or there may be volumes that resemble real works – the *Bhagavad Gita*, or *A Vindication of the Rights of Woman* – but for a single difference, in every conceivable error or combination of errors. No work could be considered reliable in such an archive, and so it becomes clear that the Universal Library would be impossible to use or even to construct. The number of volumes would be beyond comprehension, so many that they would exceed all the available space in the universe. 'No matter how we try to visualise it,' Lasswitz writes, 'we are bound to fail.'

Still, some have tried. In 'The Library of Babel', Jorge Luis Borges attempts to describe what kind of library would hold a collection like the one Lasswitz imagined. It would be, he writes, a boundless archive composed of an infinite number of hexagonal galleries separated by shafts of air. Each gallery in the library contains twenty shelves, a mirror, a closet for sleeping and a closet for ablutions, and is connected to the others by spiral staircases, which ascend and descend farther than the eye can see. The exact same number of books, with identical numbers of pages and lines per page, fill each gallery, but despite this uniformity we are told that in all the vastness of the library no two identical books may be found. Rather, its endless shelves contain 'all that it is given to express, in all languages', including 'the minutely detailed history of the future'.

Borges's library glitters in the mind. Even its surfaces – the bright volumes, shining mirrors, and polished stairs – gleam and sparkle (the title of the collection in which Lasswitz's story was published is *Traumkristalle*, or *Dream Crystals*). Yet it is also a place of nightmares, in which the endless search for knowledge becomes a tyranny that absorbs whole lives. Robert Mattaaq's humble repository pales beside Borges's terrible vision. But, although it is unlikely that either man knew it at the time, the tiny library Kpomassie discovered in Mattaaq's hut stood on the edge of a much larger archive, one of the greatest and most comprehensive ever assembled: a continuous record of eight hundred thousand years of planetary history, embedded in the Greenland ice.

The ice sheet is the product of some fairly simple ingredients – water, temperature, pressure, and time – but working together, they have the potential to remember extraordinarily precise details. Like the galleries in Borges's library, ice begins as a hexagonal flake, which fuses as it falls with other hexagons, becoming compacted with the weight of each subsequent snowfall and settling into annual layers like tree rings. The uppermost layers carry deposits of air lodged between the hexagons, but as you go deeper, much of the air is squeezed out as the snow metamorphoses first into firn (a German term that literally means 'old snow') and finally into solid ice. The pressure forces the hexagonal flakes to recrystallize, trapping a small amount of air in bubbles, which look like a swath of milk running through the ice and which can be used to build a picture of what the world was like when each layer was formed. As in Borges's library, incredible stories are spun from a relatively simple set of basic elements. Air trapped in ice bubbles provides information about the composition and concentration of Earth's atmosphere when the snow fell, and chemical traces and other physical properties of the ice reveal information about temperature, wind and snowfall that can be tagged to particular seasonal fluctuations thousands of years ago. Ice can tell stories about the frequency of forest fires and the extent of wetlands or deserts long ago and in far distant parts of the world. Ice cores can be used to date specific volcanic eruptions and even help track the past movement of Earth through the solar system. The information is highly condensed: a metre of snow may be compressed into only thirty centimetres of ice over one hundred years or so.

The Greenland ice sheet is thought to be around three million years old; the much larger ice sheet over Antarctica is more than ten times as old, but there is a limit to how far back the record can go. Even this library can't rival the infinite repositories of Borges and Lasswitz. Ice at the very bottom of the sheet, in places as much as several kilometres down, is warped by pressure and melted by heat from the bedrock, erasing much of the information and disordering the rest as if someone had torn through the archive with a pair of scissors.

Scientists began to learn to read the frozen library in the 1950s. In 1959, as Kpomassie was slowly making his way from Togo to Greenland, the US military set about building a city under the ice, 140 kilometres from its military base at Qaa-naaq. Camp Century provided accommodation for over two hundred soldiers, along with a main street, hospital, chapel, post office, laboratory, radio station, darkroom, cinema and skating rink, powered by a 1.5-kilowatt nuclear reactor, all beneath the surface of the ice sheet. A Department of Defense film from 1964 shows soldiers using electric razors and relaxing in comfortable living quarters reading *Time* magazine in their dressing gowns. The Camp Century library had four thousand volumes.

Camp Century emerged from the peculiar mix of fear and hope that shaped the Cold War imagination, partly as a bulwark against a Soviet invasion from the north, partly as a demonstration of the human potential to inhabit any environment. It was also a test site for a classified missile programme, code-named Project Iceworm ('iceworms' was also the nickname adopted by the soldiers who built and lived in Camp Century), to establish the viability of installing a tunnel network of six hundred middle-distance rockets. Excavations of the ice sheet had revealed that the ice was arranged in distinct annual layers, which puzzled the Americans, and in 1964 they handed a core sample over to a Danish scientist called Willi Dansgaard. In the early 1950s Dansgaard had developed a technique for dating the ice according to the isotopic density of different layers. He suggested that water molecules that fell during colder temperatures would contain more of the heavier oxygen-18 isotope than those that fell during the summer. The Camp Century core provided evidence that the ice was also an archive of planetary history, stretching back hundreds of thousands of years, and that this archive could be mined to create a picture of what the world had looked like in the deep past.

In 1966 Camp Century scientists and engineers became the first to drill down to the bedrock, retrieving a 1,387-metre core stratified with one hundred thousand years of snowfall. the *New York Times* called it 'the deepest and most rewarding hole ever drilled', and since then ice core scientists have pushed further and further into this record of past climates. A core retrieved from Vostok in Antarctica in 1987 revealed traces of the past two ice ages and confirmed the link between atmospheric carbon dioxide and atmospheric temperature. In the 1990s, scientists brought out a two-mile-long core from Greenland, and in 2004 the Dome C ice core drawn out of the Antarctic ice sheet described 740,000 years of atmospheric history.

But these advances in polar research also revealed just how closely the story in the ice was bound up in human history. The layers that trap atmospheric traces tell dark, and surprisingly precise, tales about us. In ice from glaciers in the Alps, a gap between layers of lead from the fourteenth century records a brief cessation in smelting, as the Black Death killed between one-third and two-thirds of the population of Europe. Ice cores from Antarctica memorialize the devastation wrought by Spanish settlers in the New World. Novel diseases decimated the indigenous populations by 90 per cent in a little over one hundred years, and the amount of land used for agriculture in the Americas decreased from sixty-two million to six million hectares, opening up huge areas as new carbon sinks. The ice record reveals an abrupt decline in atmospheric carbon of between 7 and 10 parts per million between 1492 and the onset of the industrial revolution.

Some traces braid human and geological histories together. Since the polar wastes so captivated the imagination of Romantic writers like Samuel Taylor Coleridge and Mary Shelley, the ice has often been represented as a zero landscape, a region outside time – trackless, lifeless and mindless. But the origins of some of the greatest works of the Romantic era are also recorded in the Greenland ice, in a layer of tephra – particles of volcanic rock and ash – that marks the eruption of Indonesia's Mount Tambora in 1815. The catastrophe disrupted weather patterns across the globe, casting a pall over Europe the following year that inspired Byron's 'Darkness' and Shelley's *Frankenstein*, which begins and ends on the Arctic ice. The literary critic Jonathan Bate suggests that Keats's 'To Autumn' was written in September 1819 in a burst of relief at the return of the natural order of the seasons, but something of what Keats felt can be inferred, too, in the lightening of the layers above the dark band of tephra.

Because the annual layers are so distinct, ice-core scientists can build a picture of how industrialized societies changed: not only drawing the arc of rising carbon dioxide in the atmosphere but also observing changes decade by decade, even year on year. Anthropogenic nitrogen has a distinct isotopic signature, lighter than that of nitrogen found in the atmosphere: elevated levels of this isotope in ice cores record the vast increase in oxidized nitrogen since the invention and widespread adoption of the internal combustion engine in the first half of the twentieth century, and perhaps even mark a date as precise as 1914, when the first factories using Fritz Haber and Carl Bosch's technique for converting inert atmospheric nitrogen into ammonium fertilizer began production. Layers laid down in the 1950s and '60s carry remnants of nuclear fallout and the use of leaded petrol; a shift from chlorofluorocarbons to hydrofluorocarbons marks the late 1980s, when we came to realize the damage being done to the ozone layer.

For most of our history as a species, our development has been bound up with the reach of ice, pushing us to inhabit only certain parts of the planet. The human brain evolved to its current size and complexity around two hundred thousand years ago, and with it the thinking power required to build civilizations. But all this potential was held in abeyance until the great planetary ice sheets shrank back to the poles, at the end of the last ice age, around twelve thousand years ago (itself part of a larger pattern of glaciation stretching back two and a half million years). In short order, crops were farmed, cities rose and writing began.

Ice is the planet's seat of memory. As the historian Tom Griffiths remarks, 'before Antarctica was even seen by humans, it was recording our impact'. What had seemed like a vast blank space was in fact a dynamic machine for memorialization, a global archive reaching back hundreds of thousands of years. Our traces sink into the ice, where they are locked up as if for safekeeping. The entire ice sheet, in cross section, would reveal a map of the climate memory of the planet, accurate not just to the year but to the season. And each fresh snowfall manufactures a layer of future fossils – a new deposit in the frozen library.

IT WAS A WARM AUTUMN MORNING in April, the air still and the sky hot blue, and I was standing under the looming, brick-coloured prow of the *Aurora Australis*. Four teenage girls in ball gowns stood or sat on the dockside directly beneath the ship, checking their mobile phones and makeup. Each one wore a sash emblazoned with MINI MISS TASMANIA. Their mothers hovered nearby. They all seemed indifferent to the vessel brooding above them, caught up instead in the anticipation of what looked to be a memorable day.

The *Aurora* is an icebreaker. Since it was launched from a slipway in New South Wales in 1989, it has ferried supplies and scientists to Australia's Antarctic bases during the gleaming polar summer, breaking through ice floes more than a metre thick, and it has wintered here in the Hobart docks. It was nearing retirement, though; a new icebreaker, twice the size and costing more than one billion Australian dollars to build, would replace the *Aurora* in 2020 and punch its way through the ice for the next thirty years. From the dockside, the *Aurora* didn't look like a vessel approaching the end of its serviceable life. Contrasted with the creamy white facades of nearby elegant coffee shops and seafood restaurants, it was a massive, doughty presence. Other jetties were busy with pleasure boats, but no one seemed to come or go from the icebreaker. It leaned into its own heavy stillness like a prizefighter in his corner.

The *Aurora* was docked outside the Institute for Marine and Antarctic Studies (IMAS). My family and I were in

Australia for three months while I had a sabbatical at the University of New South Wales, in Sydney, and we had travelled to Hobart so that I could give a talk at IMAS about my search for future fossils. But really, I had travelled with another purpose. I wanted to hold an ice core.

Drilling an ice core uses fairly simple technology, not unlike an apple corer. It's essentially a toothed metal pipe, spun at high speeds to cut into the ice. Cores are typically drilled in short sections and reassembled on the surface. But the stories they tell of the world as it was are astonishing in their detail and complexity. I wanted to know more about what the greatest archive of climate history on the planet can tell us about the world to come.

IMAS is home to a world-class ice core laboratory, and my hosts had kindly agreed to arrange a tour. I had left my family to explore a replica of the hut used by the Australian polar explorer Douglas Mawson. An exact copy, right down to the weather-stained exterior, had been dropped incongruously on a narrow strip of land at an undistinguished corner a few streets away. Visitors to the hut were welcomed by two pairs of fibreglass emperor penguins flanking the path like a guard of honour. The entrance to the institute, on the other hand, was watched over by a giant bust of the Norwegian explorer Roald Amundsen, who beat the British expedition led by Robert Falcon Scott to the pole.

Inside, I collected my visitor's pass from the reception desk, opposite a steel pole that rose into the atrium. Its lustreless surface blended obliquely with the building's bare surroundings, and for a moment it didn't register that this was my first encounter with what I had come so far to see. The thin spike was an ice drill, used (according to the accompanying plaque) to extract a twelve-hundred-metre core from Law Dome, near the coast of East Antarctica.

The Law Dome core was one of the few to have touched bedrock, reaching back ninety thousand years. In terms of the age of ice brought back to the surface, though, Law Dome's achievement is comparatively modest. The oldest ice retrieved from an ice core is eight hundred thousand years old, but in 2010, a team searching for blue ice in East Antarctica's Allan Hills found the first million-year-old sample. Blue ice (so called because deep compression squeezes out nearly all the air, turning the ice the colour of a summer sky) is extremely ancient; it emerges when the ice sheet is forced upwards as it flows over ridged terrain, driving the oldest ice from the bottom to the surface, where scouring winds rake away the lighter upper layers. They left the hole unfinished, and when they returned to the same spot five years later had to drill only a further twenty metres down to recover blue ice that was 2.7 million years old, frozen when modern humans had yet to evolve. Because pressure thins out the layers, and the flow of the ice can twist the deepest ones out of shape, most of the cores' information was gone, and the scientists had to date the samples by measuring traces of argon and potassium. The Law Dome core was only a fraction of the age of the Allan Hills blue ice, but because it was nearer to the surface it was the most finely legible part of the archive. The story of the last two thousand years, including the incredible leaps made since the industrial revolution, was marbled through the core extracted by the giant needle in front of me.

The difficulties involved in extracting a core intact are

immense, a level of hardship that can inspire a touch of irreverence. The water locked up in the Antarctic ice sheet is the purest in the world, and many accounts of ice core drilling mention how well it goes with spirits. *Time* reported that, during the press conference to announce that Camp Century scientists had drilled right down to the bedrock, Pentagon officials cooled their Coca-Cola with ice that fell as snow around the time of the birth of Christ. But more often, because of this combination of physical inaccessibility and fragility, the cores inspire a sense of awe. The travel writer Gavin Francis spent a year as the base-camp doctor at Halley Research Station. One summer day he received a call that some scientists at a deep-field drilling station on Berkner Island, 650 miles away, had developed a rash. Following what turned out to be a five-minute patient consultation, he was shown into the drilling tent and down a set of ice-cut steps into a blue grotto, arched like a cathedral. A hush surrounded the core itself, he writes, as if it were 'a holy of holies.'

My host at IMAS, Elle Leane, was a specialist in Antarctic literature and the only literary scholar among the institute's many palaeoclimatologists and marine biologists. We were joined by Hannah Stark, who taught English literature at the university in Hobart, and who was also interested in seeing the ice cores. Elle walked us through the building towards the lab. It was a slightly lonely existence, she said, being the only humanities scholar surrounded by scientists. The workspaces were mostly open plan, all clean surfaces and natural light, and wrapped in a monastic quiet. Most of the researchers had opted to seal themselves off from their colleagues with white earplugs. The atmosphere was cotton soft.

The ice core laboratory itself was on the top floor, and at the head of the stairs we were greeted warmly by Andrew Moy, the ice core researcher who had agreed to show us around. Once we were inside the heavy glass doors, he introduced Meredith Nation, a technician who worked with him in the lab. I couldn't help noticing that both of them were wearing shorts.

In the narrow corridor that led to the main workspaces, Andrew explained what they were looking into. 'The ice is like a library', he began. 'Each layer tells a story.'

He told us that the summer layers are lighter and airier than the darker, denser winter layers. Snow that falls in summer is subject to constant heat from the sun, which cooks the uppermost layer like a soufflé, and the ratio of oxygen isotopes preserved in the frozen water varies accordingly. Their greatest concern, he said, was contamination. Cutting, handling, and transporting the core can all introduce unwelcome influences. The ice is so pure that even breathing on it could pollute the samples. The solution is to cut a 'core' from the core: to mine the heart of the ice. Identifying the precise chemical signatures buried within it enables the researchers not only to date each layer but also to build a global picture of past climate events from the pollutants trapped in the ice and the air bubbles frozen inside it. For example, he said, lead has an isotopic signature that is unique to the site where it was mined. They could identify precisely where a wisp of lead caught in the ice had been mined and smelted.

'We could examine a sample and know from the characteristic isotope composition that the ore deposits came from

Broken Hill in New South Wales', he said. 'We read the sky in the ice.'

What are you looking for? I asked. 'We want to understand why there was a change in the heartbeat of the Earth', he replied. In the 1920s a Serbian mathematician called Milutin Milankovitć proposed that variations in the way Earth moves in its orbit around the sun would alter the distribution of solar radiation that the planet received and thus have an effect on the amount of heat energy powering the climate. According to Milankovitć's theory, the elliptical shape of Earth's orbit around the sun – its eccentricity – followed a one-hundred-thousand-year cycle. Woven into this rhythm was a counterpoint, a wobble in the obliquity of the planet as it completes each orbit, tilting back and forth on its axis every forty-one thousand years. Over the past eight hundred thousand years, Andrew explained, major ice ages have been determined by the planet's eccentricity, coming and going roughly every one hundred millennia: 'But we know from marine sediments that the pattern of climate variability before one million years ago was different, with cycles that unfolded over forty thousand years or so.' Something had forced a change in rhythm, allowing eccentricity to take over from obliquity. 'If we're going to understand the effect of what we're doing to our climate system now,' he said, 'and work out how to adapt to what is to come, we need to understand what happened eight hundred thousand years ago.'

Andrew's explanation made me think about the kind of library he worked with. Each sample might provide startlingly precise details and local tales, but to him they mattered because of what they added to our understanding of the larger story. It was as if Borges's library held just a single book, an impossibly grand epic spread across volume after volume of instalments; or perhaps the cores were fragments – a few bars, or even just a clutch of notes – in an immensely complex, multimillion-year-long polyrhythmic musical score.

A rack of fat puffer jackets was stacked to one side of the corridor. We struggled into our thermal layers and stepped up to the huge steel doors of the ice room. Once we were inside the chamber, my skull immediately began to ache. The thick cold gripped the back of my head like fingers burrowing into my brain. It was difficult to fill my lungs. The cold was like a stranger's gaze; it made you newly conscious of every exposed piece of skin. The temperature was kept at minus 20 degrees Celsius, said Meredith, who, in addition to the shorts, wasn't wearing a jacket. You must stop feeling the cold after a while, Hannah remarked. 'A bit', Meredith replied with a shrug. 'It depends how long you're in and how fast you're working.' I learned later that minus 20 is the point where nasal membranes begin to freeze.

Stainless steel tables lined the walls of the room, and the cooling system thrummed contentedly. Andrew led us to a table in the corner and lifted something the size of a small loaf of bread out of a blue cargo case.

The core seemed to reflect and absorb light at once. Cradled in my hands, it glistered in the cold strip light but was dry to touch, and beneath the glints was a kind of depthless non-colour. It glowed with an inward luminosity, a portal into an archive two miles deep. I felt as though I would sink through it into a glaucous world of cold and pressure and ancient air.

How old is this piece? I asked. 'Only about thirty-three years old', said Andrew, which startled me. It was younger than I was: elements of my own life story might be baked into this loaf-size chunk of ice. This wasn't a story to interest ice scientists, stitching together the fragments of a larger planetary narrative, but still it made me wonder about the kind of immortality ice can confer. There was an intimacy in even the possibility of holding my own history, frozen in time, which I hadn't felt in any of my other encounters with future fossils. In Borges's library there are countless copies of a single work, some that vary from the original almost beyond recognition, meaningless screeds of letters with the occasional readable line; in others the differences are so small as to escape notice – a misplaced comma or single misspelled word – such that the copies may be mistaken as identical. Perhaps this is how the ice will remember me: the volume telling my own exact story – the chemical traces I made myself – may be buried somewhere else deep in the ice sheet's infinite hexagonal galleries, but contained in this particular block there could be a history so closely like my own as to be nearly indistinguishable.

THOUSANDS OF YEARS since its demise, the Library of Alexandria persists as the symbol of the ideal archive, a place where the pursuit of knowledge was perfected. But that was not how the ancient world thought about libraries. Collections of books were mobile, migrating to new contexts where they would become the nucleus around which new collections would grow, so that it must have seemed to ancient scholars that there was only one library expanding throughout the known world.

According to the philosopher Michel Foucault, the idea of a total library, a location where knowledge is fixed, is a creation of modernity. The library as a place in which 'time never stops building up and topping its own summit', in which history is imagined as 'the menacing glaciation of the world', is, he says, a fantasy that originates in the nineteenth century, a period obsessed by the spectacle of 'the ever-accumulating past'.

It's not incidental that Foucault saw glaciers as an image of inexorable progress, but the progress he imagines is, perhaps paradoxically, towards fixity. The total library absorbs all epochs and forms but is itself immobile and unyielding. We might suppose that what gets trapped in the ice sheet is there for ever, locked in chilly immortality; that, like Foucault's archive, it is 'a place of all times that is itself outside of time and inaccessible to its ravages'. But the ice sheet is not a locked room – it is a conveyor belt. Glaciers decant into rivers and the ice sheets into the sea; even ice in the heart of Antarctica is moving slowly coastward under its own massive weight. Like the great libraries of the ancient world, the ice sheet moves and shifts, all the while remaining one and whole. As the planet warms, however, more and more ice is being lost at an alarming rate. The month after my visit to the ice core, British researchers in West Antarctica announced that a large crack in Larsen C, a fringe of ice along the continent's whiplike tail, had widened significantly, to within thirteen kilometres of the ocean. Calving was by now inevitable, it was said. When it broke free a few months later, the iceberg was nearly four times the size of London. Stories of ice loss come almost too fast to keep track. Not long after the calving of Larsen C, news came that some of the oldest, thickest Arctic ice north of Greenland had begun to break up after temperatures spiked more than 20 degrees Celsius above the seasonal average.

Most glaciers worldwide are receding. The total volume of ice in Glacier National Park is only a little more than half what it was in 1966, when the Camp Century drills first hit the Greenland bedrock. The park has lost more than 120 glaciers in the past hundred years and will likely lose the remainder by the middle of this century. The loss of ice is driven by a series of feedback loops, in which melting exacerbates melting. One such loop is a weakening of what is called the albedo effect: fresh snow reflects solar radiation away from the planet, whereas water (because it presents a darker surface, and therefore has a much lower albedo) absorbs it. In the Arctic, where the time of maximum solar radiation coincides with the beginning of summer, dull white slush and even pools of dark water gather on the surface of the ice sheet, forming deep pits, as if the ice were making cores in itself. Large drainage holes called moulins can penetrate as far as the bedrock, letting meltwater percolate throughout the sheet and ooze underneath it, easing its passage towards the sea. Soot deposits from distant forest fires and industrial activity, even in tiny grains, create pockets of heat that allow bacteria and vegetation to colonize, further darkening the ice. In fact, glaciers need not disappear completely for their data to be lost. Meltwater on the surface can percolate down through the layers, altering the chemical composition of precisely arranged strata and blurring the record irrecoverably.

According to the ice historian Mark Carey, glaciers have come to resemble an endangered species and, as with animals on the brink of extinction, there are measures to conserve at least a remnant of what was once so abundant. In 2016 UNESCO launched the Ice Memory project, an initiative to acquire and store ice samples from retreating glaciers around the world before they are gone. So far, two cores – from Col du Dôme in the French Alps and Illimani in Bolivia – have been recovered and stored in a snow cave ten metres below the surface, at the French-Italian Concordia Station in Antarctica.

However, these deposits don't come close to replicating the archive that is draining away. The IPCC estimates that, even without further climate change, between 28 and 44 per cent of glaciers worldwide will go, with far more alarming consequences than the loss of information. The ice fields of the Himalayas and the Hindu Kush, which provide water for 1.6 billion people, could lose between a third and two-thirds of their volume by 2100.

The global library of ice has become a cracked kist, a broken storehouse leaking its contents. It isn't only the chemical traces of past climates that are evaporating. There are thought to be up to a trillion pieces of plastic frozen in Arctic sea ice. Following currents that flow across the Bering and Chukchi shelves, microplastics from the Pacific that arrive in Arctic waters are gathered up by the tiny crystals that clump together to form 'frazil' (the gloss of soft ice that forms on the surface of the water) and then are bound into the sea ice. But as more of the summer ice melts, much of this plastic could find its way back to the ocean. Having long been a sink for human detritus, the Arctic is now also a source.

The seventy-four-kilometre-long Siachen glacier in Kashmir is the second largest in the world outside the polar regions; it is also, as Arundhati Roy writes, the world's highest battlefield.

Soldiers from India and Pakistan have fought on the glacier since 1984, turning it into an immense rubbish tip, 'littered', Roy says, 'with the detritus of war'. The glacier has already lost a third of its mass, however, and the battlefield is melting, spilling empty artillery shells, fuel drums and other military apparatus in a rush of recollection of twentieth-century violence. Even the Cold War's frozen legacies will not be spared the thaw. The last people to enter Camp Century, a military surveying team, did so in 1969; since then it has been left alone to endure the ice sheet's relentless embrace. Its tunnels will have thinned like unhealthy arteries, before twisting and buckling under the weight of the shifting ice, slowly leaching diesel fuel, toxic chemicals like polychlorinated biphenyls, grey water and radiological waste. A recent study suggested that this malign confection might reach the Arctic Ocean by 2090.

These human histories seeping from the ice, marred by conflict and hubris, tell a diabolical story, but the future they foretell is more sobering still. Twice as much carbon is banked in permafrost across the northern hemisphere as exists in the planet's atmosphere. As the ground thaws and softens, perhaps 10 per cent of this carbon – 140 billion tonnes – could be released over the next century, equivalent to all the carbon released to date by deforestation. Added to this is a store of methane, as much as ten thousand gigatonnes, locked up in ice crystals called clathrates, buried in the Arctic up to five hundred metres below the seabed. Methane is a greenhouse gas many times more potent than carbon dioxide, and while it lingers for only twelve years in the atmosphere, it has the potential to do much long-lasting damage. If we reach a tipping point in the thaw, then just 0.5 per cent of this methane store, released over a single decade, could increase warming by 0.6 degrees Celsius, equal to more than half of all the warming so far since the industrial revolution, driving more melting and hastening the emergence of a world without ice.

Already, the permafrost is softening, and not only long-hoarded greenhouse gases are emerging. A Siberian heat wave in 2016 exposed the remains of an anthrax-infected reindeer, releasing spores of seventy-five-year-old bacteria into the air, water, and food chain and killing a ten-year-old boy. Two years earlier researchers isolated a thirty-two-thousand-year-old 'giant virus' (so called because it is visible under a light microscope) from a sample of Siberian permafrost. The virus was still infectious (although only to single-celled *Acanthamoeba*) and was named *pithovirus sibericum*, after the Greek *pithos*, a type of large amphora like the one given to Pandora by the gods – also known as Pandora's box.

The legend of Pandora's box tells of how a host of evil was released on Earth when Pandora opened the cursed chest given to her by Zeus. But the global thaw promises a release more terrible than even the plague that emerged from Pandora's box. If the worst comes to pass, then flood will accompany drought, as meltwater raises sea levels while billions of people who rely on glaciers for water find that a source they thought would last for ever has dwindled to nothing.

IN 2018, ANTARCTIC SCIENTISTS announced a remarkable discovery: the ice was singing.

To effectively monitor changes in Antarctica's Ross Ice

Shelf from a distance, a group of seismologists placed thirty-four seismic sensors two metres deep in the firn layer, the partially compacted snow where the process of trapping air and installing its traces in the library of ice begins. The sensors revealed that harsh polar winds racing across the surface made the firn vibrate constantly, although at a frequency below human hearing. Speeding up the recording 1,200 times, however, produced an eerie keening. The scientists speculated that by listening to changes in the ice sheet's song, they could trace how quickly it was melting and even anticipate another mass calving event like Larsen C's.

For a while, after the announcement in the news of the singing ice, I found myself unable to stop listening to it. It was an odd mix of the submarine and machine-like, a low ululation very similar to the haunting soundtrack that had accompanied the video installation of a submerged New Orleans I had seen in the university's art gallery in Edinburgh. It reminded me a little of an old dial-up modem. Not long afterwards, NASA released a recording of the wind on Mars, recorded by the InSight rover. This was the first time the sound of the Martian surface had ever reached human ears, but it was not nearly as unearthly as the drone from deep within the ice sheet.

After the evil had fled from Pandora's box, left at the bottom was a thin residue of hope. Perhaps, I thought, this was it. The eerie song spoke more eloquently than any speech or report of how the ice is endangered. There is nothing like either the Ross Ice Shelf or the Mars recordings in the entire annals of human history, and yet since they were made public, they have been replicated many times over by each click, play, like and share. Who could ignore a warning like this, as the ice sang of its own dissolution? But each replay also had a small part in hastening the thaw. In just a few months the YouTube video of the singing ice had been watched more than thirty-two thousand times, and each instance involved creating an electronic trace, logged in an anonymous data centre.

In the context of the internet's astonishing capacity to record the details of our lives, the number of times the ice's song has been heard is negligible. Four million hours of content are uploaded to YouTube every hour. Every minute, Google processes 3.8 million searches, mobile phone users send 13 million text messages and subscribers to Netflix watch nearly 100,000 hours of content. Every 60 seconds, 473,000 tweets are sent and 3 million Facebook posts made, including 136,000 uploaded images. In total, social media users on just these two platforms post 645 million times per day and upload 6 billion images each month. And while the notion of the cloud might convince us that our data occupies some undefined ethereal space, floating loose of the earth, the fact is that all of it must be stored somewhere. The fabulous connectivity promised by the internet belies its earth-bound, energy-hungry reality. Every single piece of information exchanged via the internet, from government reports to the dashed-off, misspelled text apology from someone running late, is stored in one of 8.4 million data centres around the world. Our digital traces are contained not in frozen bubbles but in vast banks of hot, energy-hungry servers. Just as all plastic ever created still exists, however trivial its form and use, so a record of every idly curious web search, impulse purchase, droll tweet or snapped sunset (no filter!) persists somewhere on a computer hard drive.

Like the river of atmospheric and chemical traces that pours continuously into the ice, the internet represents an undiscriminating archive of our relationships and passions, preoccupations and whims. Indeed, for all its incredible precision, the library of ice cannot compete with the internet for the detail with which it records our lives. The block of ice I held in Andrew's laboratory could give me only an approximation of traces I have laid down, whereas the internet holds a meticulous record of my life's most insignificant moments, captured as bits of binary code and pulses of light. Or some of them, at least. Despite its prodigality, the internet's range is also profoundly narrow: 90 per cent of online data was created after 2016. Compared with the long memory of the ice sheet, patiently recording subtle shifts in the global climate for hundreds of thousands of years, the internet is a blip.

Storing this much information consumes a prodigious amount of energy, much of it wasted as heat, and also requires powerful air-conditioning systems to prevent the centres from overheating. In total, data centres account for around 3 per cent of global energy consumption per year, and while companies like Google have mostly moved to renewable sources of energy, much of the industry still lags behind, and overall is responsible for 2 per cent of global carbon emissions. Even if a time comes when the computers finally overheat, and the incredible, bewildering library we have accumulated is lost, our data will persist for thousands of years as molecules of carbon circulating in the atmosphere, wrapping the planet in a thick blanket of warmed air and contributing to the loss of more and more ice.

The irony does not end there, though. Many of the most

active internet hubs are located in coastal cities like New York, London, Amsterdam, and Tokyo: the connectivity of the web depends on the security of cities that lie either precariously close to sea level or on land reclaimed from the sea. As our drive for more data adds to the warming of the planet, the floodwaters that represent the end of one archive also threaten the elimination of another.

FOR ALL THE WONDERFUL STORIES it preserves, any library also tells a tale of loss. For every work saved, many others are forgotten. So perhaps the most tantalizing detail of Borges's fabulous library is the promise that it holds accounts of all times and places still to be. Yet the irony of 'The Library of Babel' is its completeness. A library so perfected that it closes to new acquisitions is a failure: libraries are greenhouses for new work, not simply mausoleums for the old. And when the ice goes, we lose traces of not only the world that was, but also the worlds that could have been.

No one can have loved glaciers more than John Muir. They encapsulated his irrepressible love for the beauty and mystery of wild places. But he loved them best of all because they were also the engines that made the wilderness that so entranced him. In his very first piece of published writing, an 1871 article in the *New-York Tribune*, Muir – whose works would go on to inspire the conservation of America's first national parks – described discovering a discarded book while picking flowers in the Yosemite Valley. 'Blotted and storm-beaten', he wrote, the outer pages of this book melted like the snow that had buried it, but beneath them were pages he could still read. The same condition applies to the Yosemite Valley itself, Muir stated: its 'granite pages' were likewise degraded, but 'still proclaim in splendid characters the glorious actions of their departed ice'. Glaciers weren't libraries or archives to Muir; they were pens that wrote the world into its present shape. He saw the ice as stylus, not storehouse, an instrument that covered the landscapes he loved with 'blurred sheets of glacial writing'. The ice taught Muir the still-greater lesson, as he wrote years later during a visit to Alaska, that 'the world, though made, is yet being made . . . in endless rhythm and beauty'.

Muir could not have imagined that ice would be lost on the scale we see today. Even if he was confronted by evidence of their retreat, the glaciers he witnessed must have seemed to guarantee a future of new forms and fresh vistas, albeit on a timescale he could encounter only in his imagination. But we do not enjoy the same guarantee. The last ice age peaked around twenty thousand years ago; the earth has since revolved to a low point in its irradiation cycle, with less solar radiation reaching the planet's northern latitudes. This is typically when a new glacial cycle begins, when ice begins to accumulate, slowly but inexorably, into kilometres-deep sheets of the kind that ground out the present hills and valleys of northern Europe and North America. But in the past eight hundred thousand years a new glacial cycle has never begun with more than 260 parts per million of carbon in the atmosphere, and since the industrial revolution the level has been 280 ppm and rising.

Some scientists predict that this has disrupted the rhythm of the ice cycles themselves. In 2010, a team of ice scientists performed computer simulations of three different emissions scenarios. If the total amount of carbon released into the atmosphere by human action reaches five hundred gigatonnes, the development of a new ice sheet over the northern hemisphere would be delayed by tens of thousands of years. Twice that amount of carbon delayed the next ice age still further, but if we emit three times as much – fifteen hundred gigatonnes – the simulation predicted that the next ice age could be postponed for one hundred millennia.

The palaeoclimatologist William Ruddiman has even suggested that human societies fundamentally affected the rhythm of past ice ages. According to the 'Ruddiman hypothesis', carbon dioxide levels began to rise eight thousand years ago, with methane levels also trending upwards around three thousand years later, due to a combination of human agriculture and the deforestation that accompanied it. To begin with, the changes were modest, limited to the 'slash and burn' progress of small farming groups across forested parts of southeast Europe, but the invention of the Bronze Age plow and the domestication of oxen, horses and water buffalo allowed this new form of subsistence to spread across Eurasia. Meanwhile, deforestation began in China a thousand years earlier, and irrigation techniques for rice cultivation were widespread across South-east Asia and as far as the Ganges River Valley by three thousand years ago. Just a thousand years later, Ruddiman states, almost every major contemporary food crop was already being cultivated. The surplus energy allowed populations to bloom, and where the plough led, cities soon followed. Deforestation reduced the number of carbon sinks, while rice fields filled with decomposing vegetable matter became a new source of methane emission. The net result, Ruddiman estimates, was an anomalous increase in greenhouse gases, unique in the history of the three preceding interglacials. Ice core records confirm that, for the past four hundred thousand years, these gases have risen and fallen in tandem with variations in the intensity of solar radiation, and that the ice has grown and shrunk accordingly. But between the beginning of agriculture and the industrial revolution, human activity added an extra 250 parts per billion of methane, and an additional 40 parts per million of carbon to the atmosphere, enough to forestall the beginning of a new ice age.

Ruddiman's thesis is contentious, and has invited skepticism from scientists who feel his hypothesis is ultimately unprovable. But if he is right, then the world that was formed by the retreat of the last ice sheets – *our* world, a world of art and writing, urban living and ocean travel – took the place of the one ordained by at least eight hundred thousand years of glaciations. It raises the prospect that, from its very beginning, human society has forced a rupture in the rhythm that shaped continents and governed the planetary climate for millennia. But for the extra carbon in the atmosphere, the planet would already be moving toward another glaciation. Without this break in rhythm, a very different world, of new ice-gouged lands and unseen valleys, would already be on its way.

The long process of the world's making is not stilled, of course. The return of the ice has simply been postponed, not cancelled. But even as the ice retreats, we can see a different world take shape. Temperate rainforests are growing at the former termini of some Alaskan glaciers, on earth that hasn't been exposed for thousands of years, and in Antarctica the continent's narrow fringes are greening. These new green spaces may, in time, become a source of future fossils. On their traverse of the Beardmore Glacier, en route to the pole, Scott's party discovered coal seams and fossilized plants in a sandstone escarpment under Mount Buckley, including, Scott recorded in his journal, 'a piece of coal with beautifully traced leaves in layers'. The fossil samples, which were with the frozen bodies of Scott and his men when they were discovered, were 250 million years old and proved that Antarctica was once part of the Gondwanaland supercontinent extending from the South Pole to the equator. Land now buried beneath miles of ancient ice once teemed, flocked, and swarmed with life, and as the air and seas around the continent get warmer, life is slowly returning. Already, Antarctic mosses that grew by a millimetre per year in the 1950s have tripled their growth rate, and one prediction suggests that up to seventeen thousand square kilometres of new ice-free land could appear by the end of the century.

Elsewhere, though, the contours shaped by the last glaciation will not be renewed, and valleys that would have been carved by the ice will go unmade. In the book of glaciers, even lost ice tells a story, of unwritten lands and a world that did not come to be.

WHEN I LEFT the ice core laboratory at IMAS, I headed along the marina to the MONA ferry. I had arranged to meet up with my family by the Derwent River, at the Museum of Old and New Art. Elle had said that Hobart is perhaps a surprising place to find a world-class art gallery, but MONA combines high-art seriousness with a robustly Australian slant on pretension (visitors wishing to learn more about particular works from the audio guide must press an icon labelled 'Art Wank'). The sleek lines and eccentric stylings of the museum's dedicated ferry service, painted in a grey camouflage pattern, presented a striking contrast with the *Aurora*. Passengers who wished to do so could enjoy the journey perching on one of a herd of fibreglass sheep. David Bowie's 'Life on Mars' drifted through the PA system.

As the ferry made its way upriver, past a zinc-smelting plant and a technopark, I sat on a high stool at the bow and made notes about my encounter with the ice core. On first sight of the eerie white cliffs of the Antarctic continent in 1774, Captain James Cook remarked that the unknown land before him had probably been encased in ice since the beginning of creation. But other early visitors sensed that it held hidden histories. 'And so we turned away from the mystery of the Antarctic', wrote W. G. Burn-Murdoch, who travelled with the Dundee Antarctic Expedition of 1892–93, 'with all its white-bound secrets still unread, as if we had stood before ancient volumes that told of the past and the beginning of things, and had not opened them to read.' The understanding made available by ice core science since the middle of the last century has punctured that sense of the ice as a timeless zero and opened the blind volumes to reveal an incredible record of climate and human history. But through our carelessness, those infinite pages are being torn out one by one.

Borges's librarian describes a party of fanatics who rampage through the library's galleries destroying worthless books; yet the library is so vast, he remarks with equanimity, that no human effort could ever possibly diminish it. At the end of the story, he declares that his hope lies with the inviolable archive. 'The Library will endure', he states with certainty. 'Illuminated, solitary, infinite, perfectly motionless, equipped with precious volumes, useless, incorruptible, secret'. We might adopt a similarly complacent attitude. After all, not all ice will melt. The ice locked away in the frozen centre of the Antarctic continent is protected, and it is even possible that climate change may increase snowfall over East Antarctica, leading to a gain in the amount of ice there. But complacency has proved costly before. According to legend, the Alexandrian library is said to have burned, but scholars suggest its end was more ignominious. As the influence of the Roman Empire diminished, there was simply no one to care for the fragile papyrus scrolls or make new copies. The Library of Alexandria was lost through neglect.

I arrived at MONA and caught up on how my family had spent the day, what the children had thought of the museum and Mawson's hut. Perhaps I was still seeing things through the prism of ice, but many of the artworks made me think about impermanence. There was a cascade that momentarily shaped single words in the falling water before dissolving and an installation that included a photograph of the head of Tollund Man, an Iron Age individual whose preserved body was recovered from a Jutland peat bog in 1950. An accompanying sign told me that the rest of his body had melted away after it was pulled from the bog.

We toured the exhibits until the building was about to close, and we made to leave. Taking one final turn around the Escher-like interior, I chanced upon a side room I hadn't noticed before. It was smaller than the other galleries, rectangular and about thirty feet long. In the centre were desks of pale wood, and the walls were lined floor to ceiling with gleaming white books. The room shone like quicklime.

Every cover and spine was blank; not a single page was marked.

FIVE

MEDUSA'S GAZE

Hundreds of us were caught in the bottleneck. Above our heads, helicopters circled, while a pair of bone-bright semisubmersibles trawled the shimmering water. As I guided my son and daughter through the commotion, inching our way toward the diving platform, a single thought fell through my mind like a stone: *It's there, it's right there.*

During our stay in Australia I wanted to show my children the Great Barrier Reef. Briefly, it had seemed like we might be thwarted. A storm was blowing in from paradise. In the week before our trip, a tropical cyclone had swept in from the Coral Sea, on a wave of low pressure and high sea-surface temperatures, to menace the Queensland coast. Flights in and out were cancelled, and tens of thousands of people had been evacuated in anticipation of winds of up to 270 kilometres per hour. Mercifully, the tempest weakened, and the landfall was not as devastating as predicted, diverted south at the last minute to an uninhabited stretch of coast. But severe flooding followed in its wake, swamping large parts of southern Queensland along with parts of northern New South Wales.

We experienced some of that astonishing rainfall a few days later, as we left Brisbane for Cairns, from where we would travel out to the reef. Inches of rain seemed to fall in a matter of minutes. I had never seen rain like it. Standing outside even for a moment was painful, like being caught in a storm of coins. We had all watched the cyclone's progress and were relieved that no one had been killed. In the days to come, fourteen people would be drowned in the floodwaters. But at the time the storm's passing seemed like a miraculous escape, and our thoughts turned gratefully to whether our plan to visit the reef would still be possible. All tours had been called off, as bludgeoning winds had stirred the ocean into a turbid soup.

The appointed day, however, was perhaps as perfect as could be hoped for: a blue-bathed, cloud-clear sky above pond-still water. Our destination was Agincourt, a ribbon reef on the outer system not far from Endeavour Reef, where Cook's expedition ran aground on 11 June 1770. (The botanist Joseph Banks, who accompanied Cook, records how the crew avoided a breach of the hull by dumping all available ballast, including six cannons, onto the reef.) Because of the storm it was the first trip for several days, and many of the people booked on the cancelled tours had been bumped onto ours. In all, there were over four hundred of us crammed onto the boat.

The onboard lounges thrummed with excited chatter and the chink of mugs of tea. Crew in reef-patterned uniforms buzzed through the throng, offering underwater cameras for hire and trips in the semisubmersibles. Through the PA system the captain delivered a practised stream of safety information, from advice on the advisability of wearing a stinger suit against jellyfish to the likelihood of meeting a saltwater crocodile on the outer reef (not high, but still greater than you might think). When we arrived, we would dive from a purpose-built pontoon, complete with an underwater viewing platform, and be served a buffet lunch. It was all efficient and adroit, a highly controlled encounter with one of the most poorly managed ecosystems on the planet.

The Great Barrier Reef was enduring a second mass bleaching event in as many years. More than 90 per cent of corals were affected in the first year of bleaching, and around 29 per cent of it lost; with no time to recover, it was already in the grip of another wave that some marine ecologists were calling the worst they had ever witnessed, which may even have pushed the entire system to the brink of collapse.

All this was running through my mind as we approached the reef. No doubt the tension of waiting for the cyclone to make landfall contributed, too, as well as the fact that we had only narrowly avoided the trip being cancelled altogether. But time felt short. This was an opportunity to see one of the richest and most remarkable ecosystems on the planet, one we might never get again.

Immediately after we docked on the pontoon there was a crush, as four hundred people sought the right size wetsuit, fins, and snorkel. My children fretted about the water depth. We baked in our black neoprene suits as we waited for our turn to approach the steel diving platform and enter the water.

A white-noise crash, followed by a sudden, astonishing emergence in a silent blue world.

Blunt-headed groupers slouched around, waiting lazily

for the scheduled feeding. Coal-striped zebra fish and blue damselfish like tiny electric blades came within inches of my fingers, then flicked away. Day-Glo parrot fish bobbed and swerved around the ungainly flop of our neon flippers. As my son turned smooth somersaults in the water like a porpoise, I could see the corals below us had a piebald mottling: large patches gleamed ossuary-white among the warm browns and the supercharged blues and pinks.

Suddenly, my daughter dropped her plastic snorkel. I tried to grab it and missed, and could only watch as it drifted slowly out of reach. Peering closer, I saw there were other lost snorkels lying on the sandy floor nearby. Each one glowed eerily, like broken stalks of living coral.

WE HAD BEEN ABSORBING grim predictions about the fate of the reef since we had arrived in Sydney, in the middle of a 40-degree-Celsius heat wave. But not everyone seemed concerned. The other big news story was the proposal for a new giant coal mine in the Galilee Basin in southern Queensland, which would export sixty million tonnes of coal over the next sixty years. Despite loud warnings from scientists and activists, a delirium seemed to have taken hold of the decision-makers. The week after we arrived, Scott Morrison, an Australian MP who would later be elected prime minister, brandished a large fist of coal on the floor of the parliament. 'This is coal', he said fatuously. 'Don't be afraid.'

The Great Barrier Reef isn't, in fact, one reef, but a long arc of more than thirty-six thousand individual reefs, stretching twenty-three hundred kilometres from its northern tip to its southern extremity. In the north it presents a narrow, semi-continuous barrier that opens out to a broader scattering of patch reefs in the south. What looks like a single structure is in fact many smaller reefs, and the same disaggregation occurs when we look at the reef through time. It's one of the planet's oldest continuous ecosystems, but the current Great Barrier Reef is only the most recent in a series of incarnations, succeeding generations stacked one on top of the other like ancient civilizations. The first layers were laid down perhaps half a million years ago, but the uppermost – the modern reef – began to form around nine thousand years ago. Coral polyps are the planet's builders, continually constructing new structures on the fossils of older corals. The reef is the only living thing that can be seen from space, and the largest living structure on Earth; it's also home to an astonishing abundance of life. Fifteen hundred species of fish and four thousand species of mollusc, as well as hundreds of different kinds of birds and more than thirty whale species, rely on the reef for shelter, food, and space to reproduce. Worldwide, 25 per cent of all marine life depends on coral reef systems that, all told, occupy only 0.1 per cent of the ocean floor. But these oases are under severe threat. Corals around the world are dying: bleached by rising temperatures, drowning in rising seas, and, as the oceans are packed with more carbonic acid, starved of the calcium carbonate they need to build their underwater cities.

A coral polyp is a tiny, soft-bodied creature – essentially, a stomach bulb topped by a crown of delicate, fingery tentacles – reliant for nutrition on a symbiotic relationship with photosynthesizing algae called zooxanthellae. The algae also provide corals with their vivid pigmentation, from honey brown or peach to a lysergic array of flamboyant yellows, blues, and pinks. When the water gets too warm, however, the coral expels its symbionts, vomiting out the algae and, with it the spectacular colours, leaving behind a stripped and starving skeleton (literally: without their symbionts, the corals starve). Dying corals glow with a sepulchral whiteness before fading to a dead, ashen grey.

As well as the bleaching effect of warmer seas, they are vulnerable to changes in ocean chemistry. A third of the carbon dioxide produced by burning fossils fuels since the mid-nineteenth century has been absorbed by the oceans – somewhere in the region of 120 billion tonnes. Dissolved in water, carbon dioxide produces carbonic acid. Already, the pH of the world's oceans has declined by 0.1, a seemingly modest figure that masks a far graver reality. As pH is measured on a logarithmic scale, 0.1 represents a rise of as much as 30 per cent in acidity. Corals need calcium carbonate to build their structures; crustaceans, including krill at the very bottom of the marine food chain, need it to build their shells. More carbonic acid dissolves more carbonate ions in the water, so the reef and shell builders simply don't have enough materials to work with. This change is effectively irreversible, at least on a human timescale. It will take tens of thousands of years for the seas to return to their pre-industrial chemical state.

Corals also face other threats. The polyps must build continually to keep their coral structures near enough to the surface to receive light. If rising sea levels sink a reef below the photovoltaic zone, its zooxanthellae can't photosynthesize, and the polyps will starve: in other words, coral reefs can drown. They can also be subject to waterborne diseases that circulate more readily in warmer waters. Tropical storms, which are becoming more common and more violent, can also damage vulnerable reefs: one of the marine scientists on board our tour told us that cyclones like Debbie rolled over the surface-lying corals with the force of a bulldozer.

I felt keenly aware that there was something morbid about our trip to the reef: *see it before it's gone* has become an all-too-common justification for visiting vulnerable ecosystems. But seeing can be a necessary step towards believing. I wanted my children to see one of the world's wonders, but I also wanted them to appreciate first hand how fragile it had become. And for myself, perhaps most ghoulishly of all, I wanted to see a future fossil in the making. Most of the future fossils I have encountered in my search will take many hundreds of human lifetimes to form. The rate of sea level rise is slow enough that most coastal cities will have centuries to adapt; the uncountable objects we bury in landfills will be unchanged for decades. But unless there is radical action to address the harm being done to the world's oceans, most coral will be dead within a generation. In our lifetimes, we may see it transformed from the world's largest living ecosystem to a mountain of dead stone.

ON THE REEF TOUR we were warned not to touch the corals, as human contact might cause damage, transferring harmful bacteria or removing important algae. Coral is also razor-sharp, and wounds can lead to infection and even blood poisoning.

Indeed, throughout history this beguiling amalgamation of living creature and inert rock has invaded our imaginations with feverish visions and fabulous dreams.

The ancient Greeks thought corals were plants that petrified on contact with the air. In Ovid's *Metamorphoses*, the hero Perseus rests on the shore, having defeated a sea monster that threatened to devour his bride, Andromeda. Next to him lies a trophy of an earlier victory: the head of the Gorgon Medusa, the sight of which turns any living thing to stone. As he washes the serpent's blood from his hands, the plants around the Gorgon's head of snakes begin to harden into rock. 'Even today coral retains this same nature', Ovid writes, 'hardening at the touch of air.'

Coral seems to have had a hallucinatory, shape-shifting quality for later writers too. 'Of his bones are coral made', sings Ariel in Shakespeare's *The Tempest*, of the supposedly drowned King of Naples: 'Nothing of him that doth fade, / But doth suffer a sea-change / Into something rich and strange'. In 1646, the English antiquarian Thomas Browne cast doubt on the assumption 'that Corall is soft under water, but hardeneth in the ayre', and proposed that the experiments of Boethius, who handled coral a fathom underwater, indicate that its 'concretion' was due to 'the coagulating spirits of salt, and lapidificall juyce of the Sea'. Browne reports that a man sent to dive down a hundred fathoms (an implausible 180 metres) to observe whether the coral was hard or soft returned bearing in each hand a branch of coral 'as hard at the bottome, as in the ayre where he delivered it'. Still, the perception that corals were plants in water and rock outside it persisted until Charles Darwin deduced that enormous reef structures – 'mountains of stone' that surpassed even 'the vast dimensions of the pyramids' – were in fact the work of tiny polyps. Coral fascinated Darwin, for whom the prospect of visiting coral reefs was the sole consolation for enduring the miseries of the *Beagle* voyage, the violence of Pacific storms, and his chronic seasickness. 'I hate every wave of the ocean', he declared, but even the thought of coral was 'enough to make one wild with delight'.

Corals are a source of delight, too, in Paul Klee's dreamlike painting *Sunken Landscape*. The picture is a riot of superabundant colour. Branching bloodred and chlorophyll-green structures wave and dance, tined and curlicued to resemble the reef gardens I saw at Agincourt with my children. It's a fantasy of life inverted: there's even a great big upside-down daisy suspended from the top edge like a floral sun. It's a joyful scene, surging with life – except in one detail. Like a dark twin to the sun-daisy, a black sun hangs, slightly decentred, in the middle ground of the picture. In *Chroma*, a colour-memoir he wrote as AIDS-related illness was blinding him, Derek Jarman wrote of how each colour possessed its own sense of time. 'Passing centuries are evergreen', Jarman observed. 'Red explodes and consumes itself. Blue is infinite'. But lurking 'behind the blue sky' there is 'a black without end'. Despite the clamouring hues and cavorting forms, when I look at Klee's picture it is this dense circle that always holds my eye. It seems to pull all the surrounding light and colour into itself, a vortex for my attention.

I was in my early twenties, studying for my master's degree, when I first saw *Sunken Landscape*. My class was reading *Omeros*, Derek Walcott's Caribbean version of *The Iliad* and the *Odyssey*. It is a monumental poem in every respect, washed by epic tides of classical poetry and the coruscating history of the Atlantic slave trade. In an early section, Walcott's modern Homeric hero Achille dives for conch shells to sell, illegally, to tourists. He ties a concrete block to his ankle and falls into a soundless world populated by the coralline bones of his murdered forebears, dumped like so much unwanted ballast from slave ships during the Middle Passage. He feels his skin begin to calcify. Later, afflicted by sunstroke while fishing in deep water twenty miles off the coast, Achille hallucinates a repetition of the Middle Passage, walking to the Caribbean from Africa across the Atlantic seafloor, 'through vast meadows of coral' like 'huge cemeteries of bone'.

Today, the metaphor has become a reality; the meadows have themselves become graveyards. The coral groves of Walcott's imagination have suffered colossal losses in the past fifty years, declining by an average of 50 per cent; in extreme cases, some Caribbean reefs have been reduced to only 10 per cent of their former extent. Overfishing, excessive coastal development, pollution, disease, hurricanes, and a series of bleaching events have all played a part. Even the deeper-lying corals, below thirty metres, are vulnerable to sedimentation and damage by bottom-trawling fishing nets. The majority of staghorn corals in the Caribbean have been lost, and the stony ruins are already being covered by a layer of mud.

As we discussed this passage in class, our tutor directed our attention to a reproduction of Klee's painting on the wall. 'When I read this part of the poem', he said, 'I always imagine I'm inside this painting.' I remembered his words and the painting when we were diving at Agincourt, and felt that same sensation of moving through the world of colour Klee had imagined. But it wasn't until I returned from our visit to the reef and looked at the picture again that I noticed the white patch of coral in the foreground, looming out of the shadows under that coal-black submarine sun. I knew it was fanciful to suppose that Klee had prophesied that our addiction to fossil fuels would bring about the reef's destruction, but still I couldn't look away. His painting seemed to me, now, like an uncanny vision, one that held me transfixed.

THE CLASSICIST Jane Ellen Harrison regarded the Gorgon as a kind of evil eye. 'It slew by the eye', she writes, 'it *fascinated*'. Those who looked upon it were captivated by its gaze and turned to stone.

According to Ovid, Perseus was conceived in a shower of golden rain. Returning on winged sandals to the island of Seriphos with the head of Medusa, he was 'driven by warring winds over the vast expanse of sky: like a raincloud, he was blown this way and that.' Perseus searched the face of the earth for a place of refuge until, in fading light, he came to the kingdom of the Titan Atlas, who, Ovid tells us, 'surpassed all mortal men in size'. The Titan was also paranoid, haunted by a prophecy that one day a visitor would rob him of his precious orchards. Perseus begged for shelter, but the truculent Atlas refused. Unable to beat him in a struggle, Perseus brandished the Gorgon's head. Instantly, 'Atlas was changed into a mountain as huge as the giant he had been.' His bones became stone, his head and shoulders an immense ridge bearing upon it all the heavens' riding stars.

It is thought that corals were named *gorgia* in Greek by

Metrodorus of Scepsis, after Gorgias, a Thessalonian orator who lived to the age of 109, because both seemed to have been petrified by age. The association persisted: Linnaeus called horn corals *gorgonia*. But the Gorgon is an ancient figure in Greek mythology, present in primitive rituals that precede Ovid's elegant Roman narrative by hundreds of years. In Greek art, the *gorgoneion* (the name for the Gorgon emblem that decorated Greek shields) is unique for appearing full-face rather than in profile, a grotesque mask with a leering mouth, waggling tongue, and fatal stare.

Perhaps the most famous modern *gorgoneion* is Klee's *Angelus Novus*, a watercolour sketch painted in 1920. It depicts an abstract angel in full face against a pale background, with bulging eyes and a gaping mouth of knifelike teeth, its hair a waving mass of scrolling forms, arms raised and pointing to the sky. The pupils in its staring eyes are a deeply scored black. In 1921, Walter Benjamin bought Klee's painting for a thousand marks. He kept it in his home for the next twelve years, until he fled Nazi Germany in 1933. In spring 1940 (the same year Klee died), only months before Benjamin took his own life on the Spanish border, the angel returned to haunt his last surviving piece of writing. His brilliant and elliptical 'Theses on the Philosophy of History' pictures Klee's angel as 'the angel of history', mesmerized by the calamitous spectacle of history piling its wreckage at his feet. Benjamin possessed what Theodor Adorno called a 'Medusan glance', by which everything he surveyed was turned, alchemically, into the stuff of myth. *Starren*, in the original German, means to stare, to observe intently, but it can also mean to stiffen or petrify. The angel is petrified by the catastrophe unfolding before him, but his gaze also petrifies those who look upon him. We can't help searching it for traces of what Benjamin saw in its gaping expression, traces of the future to which our backs are turned, while the monuments to progress are already piling up in heaps of broken stone.

As Perseus' eye is drawn to the brimming abundance of Atlas' orchards, the Great Barrier Reef fascinates us with its scale and richness; like Atlas, the reef, once dead, will leave behind a monumental body that will endure for hundreds of thousands of years. Because of its immense size, the dead reef will remain as a memorial in stone to lost biodiversity. Future visitors will still be able to swim over the lifeless reef just as we did, although they will almost certainly be deeper underwater. Astronauts will see it from orbit. In time, much of the substrate will be eroded in the more acidic water, but much will be preserved, as the rising seas that will entomb cities like Shanghai and New York also cover the dead corals in thick sediment, just as has already begun to happen in the Caribbean. With the polyps no longer erecting their ever-rising cities, the ruins will sink beneath the mud. When the reef can no longer be seen on the surface, perhaps only a hundred years from now, scientists may visit the site to take sediment cores, plunging through the layer of mud to recover creamy spears of dead coral, from which they will learn about how coral cities once supported so much life in the oceans and about what killed them.

In time, if oceanic conditions recover, reefs may be established again on the site of the Great Barrier Reef, just as it was established on the ruins of older reef communities. If so, then many millions of years from now the claggy layer of sediment that coated and preserved the dead reef will itself show up as a reef gap – a geological boundary between bands of calcium carbonate. We can read a similar gap in the ocean cores that mark the Palaeocene-Eocene Thermal Maximum (PETM) when, fifty-five million years ago, the global mean temperature rose by more than 8 degrees Celsius. The oceans absorbed a huge quantity of additional carbon, acidifying rapidly. Calcifying organisms vanished from the fossil record, their absence marked by a deposit of red-brown clay. Millions of years from now, any future geologists will be able to read a similar smear of red, like a blush of shame, marking the disappearance of coral and the thousands of species that depend upon it.

Some coral will persist, most likely cold-water species such as those that grow in the northern hemisphere. Tropical regions might see coral grow again at some point in the planet's future, but they will do so in a world as different from what we know today as the parched continents of the Early Eocene. Until then, unless we make drastic changes in how we manage the oceans, the Great Barrier Reef will stand as an enormous future fossil, 2,300 kilometres long. Might there still be tourism to the reef when it's dead, a kind of pilgrimage to mark what was lost? Will our descendants visit the deserts of these former marine oases, lamenting our unwillingness to act? Perhaps they will only be able to avert their eyes, just as we have done.

As I learned more about the future of coral reefs, I felt that black sun in Klee's painting beating down upon my brain.

BEFORE OUR VISIT TO the reef, I had attended a workshop at the University of Sydney organized by Iain McCalman, a historian who had written a history of the reef. He knew about my interest and had suggested that I come along. 'I'll introduce you to Jody', he said. 'You can see some of his coral cores.'

Jody Webster is a sedimentologist who studies the palaeo-climate of coral reefs. I found out in the bar afterward that he spends some of his spare time exploring Sydney's storm drains, looking for what he called 'drain stalactites' – accumulations of calcium just like you'd find in a natural cave system. He'd brought three cores with him, taken from different parts of the Great Barrier Reef. They each represented a distinct period in its history. The oldest was more than 125,000 years old, the remnant of a previous system that had died before the last ice age began, when the entire reef system was almost drowned by rising seas. Extracting it had cost something like twelve million Australian dollars, Jody told us. The next sample was fifteen thousand years old, a bit of 'pre-Holocene' reef. The most recent had grown and died in our lifetime.

We passed the cores around the room as he told us about how they were extracted from an isolated patch of the southern reef, and how each core sample told a distinct story about the history of this ancient ecosystem. Cut into palm-size discs, each sample had a buttery warmth. They could have been sculpted in cream, but the surfaces were pumice-coarse to touch. The prints of tiny fossilized shells were embedded in the older cores. I hefted each one in my hand. They had a pleasing density, like paperweights.

During a break in the workshop, I'd told Jody about my interest in future fossils, and he invited me to visit him in his lab to see some of the larger core samples. These palaeo-reef traces offered a snapshot of past climates and reef environments; I hoped that they might also help me imagine the kind of future that fossil coral would make in a world of febrile seas.

A few weeks after diving at Agincourt, I knocked on the door to Jody's office at the University of Sydney. He had himself just returned from One Tree, a small coral cay in the southern reef system where the university has a research station. As I sat down, he opened a file of photos on his desktop monitor to show me what he'd found out there and began scrolling through dozens of high-resolution, close-up images of the coral. The effect was oddly sedating – it took me back to the serene experience of drifting above gardens of blossoming colour – but large patches of the photos gleamed white, marbling the images like globs of fat. When I had first met Iain, he'd encouraged me to get to One Tree, where the coral was relatively pristine. But Jody's photos showed that bleaching had occurred there too. 'It's the first time I've seen this level of bleaching in the eight years I've been coming to this part of the reef', he said. Older scientists at One Tree had claimed they hadn't seen anything like it in twenty-five years.

I told him about Agincourt and how, although swimming among the coral had been sublime, the frantic pace and the insistent noise of the helicopter flights had struck me almost as much as the reef itself. The sense of emergency seemed somehow closer to the real state of things than did the idea of some tranquil sea garden, I ventured. But to my surprise, Jody was having none of it.

'We can't give in', he insisted. 'However bad it is in places, there is still a lot of healthy coral. In any case', he said, 'we don't have a choice.'

Jody was due to set out for another research trip the following day, visiting a series of drowned fossil reef sites along the entire length of the reef, from Fraser Island on the south Queensland coast to Torres Strait in the north. Their object, he said, was to find evidence of changes in temperature and in the behaviour of the East Australian current over the past one hundred thousand years. 'Knowing what killed the reef in the past can help us to understand what is killing it today', he said. 'We're detectives: everything – sediment flux, water chemistry, temperature – is a suspect.'

We moved to a couple of armchairs in the corner of the office, and he began to explain how indigenous people living along what is now the Queensland coast had coexisted with the reef through periods of climate change for thousands of years. He sketched a diagram on a whiteboard. 'Sea level has been stable since around seven thousand years ago', he said. But during the Last Glacial Maximum, about twenty-one thousand years ago, it was 120 metres below what it is today. The shoreline extended in places to the reef's current position at the edge of the continental shelf. What is now a shallow sea, sheltered by the outer reef, was then a habitable plain. The inundation that flooded it as the ice sheets melted also submerged the complex valleys of what is now the Parramatta River, creating the folded inlets of Sydney Harbour.

'This time is still remembered in some indigenous cultures', Jody said.

I wrote down what he told me, although it seemed incredible that a vanished landscape could remain preserved in language for so long. Later, I looked into this startling possibility. In general, linguists suggest that cultural memories can't persist beyond five hundred years, eight hundred at most, without the 'core' information being lost beneath layers of later embellishments, like those that have hidden the historical authors of *The Odyssey* from us. But there are numerous stories among the indigenous groups around the Queensland coast that recount the time when the shoreline lay 'where the barrier reef now stands'. A Gungganyji story remembers when a man called Gunya, who lived near Yarrabah, now Cape Grafton, ate a forbidden fish. This angered the gods, who caused the sea to rise and drown the land. Gunya and his family escaped to higher ground, but the sea never retreated to its former limit. Memories of this time also seem to be sedimented in language. The Yidindji name for Fitzroy Island is *Gabar*, meaning 'lower arm', which suggests it was once a promontory; the word for island, *djaraway*, means 'small hill'. The stretch of water between Fitzroy and King Beach is called *mudaga*, the name for the pencil cedars that once must have grown there. These stories and place names represent an astonishing feat of geographical memory, recalling a landscape that hasn't existed for at least seven thousand years.

Metrodorus, who named corals after the Gorgon, was famed for his fabulous powers of memory. But there is nothing in classical or indeed any other world literature that can compare with the memory feats performed by the Gungganyji and Yidindji peoples. Their stories and place names devastate the conventional wisdom that, without writing, language is a poor, leaky receptacle in which to carry memories into the deep future. They raise the prospect that our descendants, in turn, might dream epic fables to explain the dramatic changes in sea level rise and the presence of strange city-islands rotting in the shallow seas that will cover what was once our coastline. Perhaps the Great Barrier Reef will persist as a fossil presence in the language and legends of the distant future, torqued by the pressure of time into new mythic shapes but bearing the essential imprint of a grand, once-living structure.

Jody's day was crowded with meetings before he was to join the research ship in New Zealand, so Madhavi Patterson, one of Jody's PhD students, had agreed to give me a tour of the lab. As they escorted me through the quiet halls, Madhavi told me about her research, which involves looking for clues about the palaeo-environmental conditions that shaped earlier incarnations of the reef. There were various chemical methods to establish past biochemical changes, but also more tactile means: handling the samples, feeling for changes in texture as well as pattern and colour. 'Reading the coral', she called it. Cores can show visible evidence of stress events. At the workshop weeks earlier, Jody had compared the coral samples to layer cakes. Much of the work involves searching for terrigenous layers – soil horizons, plant matter – that record when a reef died and was covered by sediment. Some cores even show microscopic roots.

We paused at the door of the laboratory. 'We're looking for "other environments" embedded in the coral', Jody said before he left us.

The coral cores were kept in a long, rectangular room with benches running the length of the far wall; the centre of the room was dominated by a huge table covered in hundreds of sample trays. Perhaps it was the earlier talk of detective stories and searching for clues, but my first thought was that it looked like a crime lab, with bodies laid out for examination.

Some were perfect cylinders, luminous as church candles; others seemed to have crumbled into fragments and looked like broken plaster or uncooked dough. Yet each tray was meticulously labelled, and the orientation of the sample carefully noted with a lipstick-red arrow indicating which way was up.

Madhavi handed me one of the cylinders. It was a sample of brain coral, she said, probably fifty years old when it died 120,000 years ago. It was longer than my arm, and I worried I might drop it. The texture was rough but pleasing, the shaft inscribed with a fine latticework tracery. It certainly felt dense, but it was also lighter than I expected, like handling a large bolt of lace. Even the sound it made as it slid in and out of the sample tray was appealing – a resonant, almost musical scraping, as if the core were a tine on an outsize tuning fork. In some of the other samples I could see branching patterns, even the imprints of shells, a braille that told of the life that had constructed the coral and that had lived upon it. For all their heft, the samples seemed fragile. Even though I handled the core as gently as I could, tiny fragments came away on my hand. I noticed later, as I left the lab, that my shirt was coated in a fine white coral dust.

Madhavi told me about a discovery she'd made on a recent trip to One Tree: a glass Coke bottle cemented in the coral flats. It had probably been there for around five years, she said. The label had long been stripped away, but the brand name on its side and the distinctive shape left no doubt about what it was. Dropped casually overboard one day, the bottle was slowly being interred by the reef – cast in stone.

Another research student, Belinda Dechnik, was busy

in the corner examining samples with a microscope. 'These are from Western Australia,' she said, 'not the Great Barrier Reef. There's no protection for coral out there. They're drilling directly into it, looking for oil.' She invited me to take a look at what she was examining and made room at the bench.

I'm not used to looking down a microscope, and it took a moment to focus; when I did, it seemed like I'd fallen into a lunar landscape, pitted with the impacts of tiny meteors. With each slide, a new dreamscape of pattern and colour bloomed before my eyes, like the mottling on degraded film stock. One slide was filled with orange pictograms, like Aztec sunstones. I gazed on, fascinated.

'These icons, they're incredible', I said, finally looking up from the microscope. 'What are they?'

'Those are forams', Bel explained. 'Foraminifera – microscopic fossilized organisms layered in the reef sediment.'

She switched the samples again, this time showing dead corals. 'These were gathered from the seabed by a drilling robot, eighty metres below the surface', she told me. Whereas the other samples had been captivating, these were oddly repellent, covered in irregular brown mottling. They looked like slices of mouldy bread. She told me about a recent trip she'd made to sample coral reefs in South America. There are corals growing at the mouth of the Amazon in conditions that would wipe out the corals on the Great Barrier Reef.

'You're swimming in water that's too muddy to see anything', she said, 'and yet there's these corals!' But even there, she had found evidence of bleaching.

I went back to the coral cores and took some photos. As I peered closer, new details revealed themselves. One unprepossessing rhomboid chunk had a milky swathe of leaf-prints running down the middle of it; to one side, a cone shell in cross section revealed its fluted columella, the shell's central pillar, the plaits twisting around it like ivy. I thought of the terrigenous layers that spoke to coral scientists of calamitous change: signs of life that were also markers of coral death. There were more fingerprint whorls; one sample bore what looked like a sine-wave pattern, another a ribbon of tracks like a bird's footprints. The column of brain coral I'd held was wrapped in cascading lines of pinpricks like hundreds of minuscule mouths, gaping in petrified outrage.

BENJAMIN'S BRIEF ACCOUNT of Klee's angel is prefaced with a line from a poem by his friend Gerhard Scholem. *'Ich kehrte gern zurück'*, it reads: 'I would like to turn back.'

Around the world, a wide variety of projects aim to turn back the dire prospects faced by coral reefs. Activities like coral farming, where fragments of reef are collected and cultivated in special aquatic nurseries, and Rigs to Reefs programmes, which repurpose decommissioned oil platforms as homes for displaced marine life, offer hope that at least some of the amazing diversity of coral ecosystems might be preserved.

During the workshop at the University of Sydney where I'd first met Jody and held the coral core samples, I also met Renata Ferrari and Will Figueira, who had designed a method for 3-D-printing artificial reef systems. Most replica reefs fail to some degree because they're too symmetrical. The cross-hatched legs of a redundant oil rig or a stack of breeze blocks can't compete with the intricacy of coral. But Renata and Will made prints from 3-D maps of real reef structures, meaning they could recreate in plastic the exact shapes and textures of a vanished reef. With enough resources, they could map the entire length of the Great Barrier Reef and store it on file, preserving a digital imprint of the reef even if the real one is lost.

Renata and Will showed us a computer-generated visualization of the reefs they proposed to print. It was a section of coral about a metre square, as distinctively ridged and spiny as the real thing. By manipulating the image, they could give us a 360-degree vista on the digital reef, flipping and spinning it like pizza dough. They had also brought some real samples of 3-D-printed coral, and just as Jody had done, they passed their samples around the room for us all to hold. The likeness was uncanny, from the coarse-grained surfaces to their dead white colour. Unlike the real coral, however, the plastic replicas seemed almost feather light, as if they might simply float away.

Roland Barthes declared that, despite its capacity to mimic, plastic is a disgraced material: whatever final shape it assumed, he said, it would never attain 'the triumphant sleekness of nature'. Renata and Will's printed reef seemed to dispute this, replicating every kink and node of the existing reef structure in a material that, unlike coral, would be indifferent to the acidity or temperature of the future oceans. But if we have to remake twenty-three thousand kilometres of dead reef from plastic, the disgrace will surely be ours.

Advocates of artificial reefs know that they aren't a permanent solution to the problem of warming oceans, but they hope that the technology involved might buy enough time for coral to hold on while we address the causes of warming and acidification. But as the chemistry of the oceans changes, it becomes less likely that there will be anywhere in them for coral to survive at all. Ultimately, the real benefit of Renata and Will's invention could simply be as a feat of memory: a digital archive of lost riches, detailed with a precision to rival that imagined by Borges in 'The Library of Babel'; or perhaps a better analogue would be the mapmakers in another of Borges's fables. 'On Exactitude in Science', a single-paragraph fragment, describes an empire whose cartographers can produce maps that render their subjects at a scale of 1:1. Mapped city matches real city stone for stone; the map of the empire itself covers an area exactly equivalent to its subject. But eventually, Borges says, the art declines, and the map of the empire is left to become a 'tattered ruin' in the 'Deserts of the West'.

There are other reasons to hope, however. The bleaching I witnessed at Agincourt turned out not to have been as severe as that of the year before, even though temperatures were higher and despite the ravages Jody had witnessed at One Tree. Marine biologists studying the effects of the consecutive bleaching events found that the first wave left behind a kind of geographical footprint that moderated the effects of the second event – whereas coral exposed to sustained temperature spikes of 4 to 5 degrees Celsius in the first year had a 50 per cent chance of bleaching, in the following year half the corals exposed to a spike of 8 or 9 degrees Celsius survived. Areas previously exposed to heat stress proved more resilient: the extent of the second bleaching, they surmised, was contingent on the severity of the first. The reef was demonstrating a form of ecological memory, 'recalling' past times of heat in order to better cope with future warming.

Despite this welcome discovery, the scientists were cautious. One likely mechanism for the improved resilience was the mass mortality of more heat-sensitive corals – rather than remembering how to survive in warmer waters, the reef seemed to be forgetting those parts of itself that could not stand the heat. Even if some species were showing a capacity to acclimatize, they said, the coral would be unlikely to keep pace with the rate at which the oceans are changing or the scale and frequency of future bleachings.

Digital mapping may allow us to remember lost reefs; coral itself may even mimic its namesake Metrodorus, 'recalling' the past in order to adapt to stress in the present. But neither alone will provide a viable future for coral reefs. Perseus found victory over Medusa with a polished shield in which he saw both his own image and the monster caught in the same reflection. It isn't feats of memory that will save coral, but a willingness to turn and face the damage we have done.

<hr>

SIX

THE MOMENT UNDER
THE MOMENT

As I turned off the highway, a sign by the side of the road offered a stern warning to stay on the bitumen and in my car. Also, to look out for kangaroos.

This seemed like good advice. Marauding marsupials aside, the wet season had yet to recede. Creeks and rivers were still in flood, and much of the surrounding bush was underwater, providing temporary avenues for saltwater crocodiles to patrol. Other than Jabiru, a town a few kilometres to the west with a tiny airport and a population of just over a thousand people, there was little but bush in any direction for hundreds of miles. The afternoon was beginning to taper towards evening, but it was still hot and humid, and the sky above Kakadu had a brassy sheen that hurt my eyes.

Kakadu, a national park in Australia's Northern Territory, is truly vast: nearly twenty thousand square kilometres encompassing mangroves and mud flats, tropical rainforest and glowing sandstone cliffs. Some of the oldest rocks on the planet are here, 2.5-billion-year-old granite intrusions that have lingered at the surface since Earth was half as old as it is now. Others are imprinted with ripple marks left by the immense river that laid down the sandstone 1.7 billion years ago. According to the indigenous traditional owners, the landscape was created by the passage of the Rainbow Serpent, the oldest of the ancestral spirits, who conjured the escarpment and surrounding floodplains with her billowing song and undulating body.

The Waanyi author Alexis Wright's epic novel *Carpentaria* opens with an account of 'a creature larger than storm clouds . . . laden with its own creative enormity', who drew her huge body across the territories that cup the Gulf of Carpentaria like hands around a great bowl, threading the earth with rivers and gouging deep valleys. When her work was done, Wright says, the serpent descended into the limestone under the Cape York Peninsula, where she dwells in a maze of aquifers, her breath to this day dictating the rhythm of the tides and the seasons.

The Jawoyn people tell of how the southern part of Kakadu was made by Bula, the lightning spirit. During the Dreaming, the time of creation when the world was in flux, Bula arrived from the saltwater country by the Timor Sea in the company of his two wives, looking for things to hunt, and such was the vigour of his hunting that he reshaped the entire territory. Finally Bula descended, like the Rainbow Serpent, into the earth, where his body was transformed into minerals. It is said that the gold veining the hills is his essence, the residue of his lifeblood, but there are also high concentrations of toxic heavy metals such as arsenic, mercury, and lead at sites associated with Bula. These places are *djang andja-mun* – sacred, or dangerous – and must not be tampered with. Bula guards his sleep jealously. The Jawoyn warn that if Bula is disturbed, terror will break over the land. Those who approached the sacred sites would sometimes be struck with strange afflictions. In recognition they called it Buladjang, or Sickness Country.

Sickness Country extends throughout Kakadu, but although I was surrounded by thousands of square miles of bush, where I was going wasn't actually part of the park. At the end of the road was Ranger, an open-cast mine that is the source of nearly 10 per cent of the world's uranium. Kakadu was established as a national park in 1978, but the mine and nearby Jabiru were exempted, creating islands of industrial activity in an ocean of tropical wetland and stone country.

As I dodged potholes I remembered the last time I'd been here, twelve years earlier. It was the wet season then, too, and my wife and I had travelled by boat up drowned roads to Ubirr, one of the most arresting rock-art sites in northern Australia. Fred, our guide, walked us through galleries that had first been painted forty thousand years ago by the ancestors of the Mirrar Gudjeihmi traditional owners. There were X-ray-style paintings of fish and animals, and – made more recently – thick-bodied Europeans with their hands thrust in the pockets of their bulky trousers; high up on an overhanging rock, perhaps twenty feet above the ground, a ghostly white thylacine was clearly visible. Thylacines, or Tasmanian tigers, have been extinct on the mainland of Australia for at least two thousand years.

How did the artist reach up there? I asked Fred. 'Well, some people say there must have been a tree nearby', he replied. 'But we reckon the artist was a shaman, and he used magic to fly up. Or maybe he pulled the rocks down to him!'

One figure, etched in red, seemed to writhe in pain. Its splayed fingers clawed the surrounding rock. The joints of its arms and legs were massively swollen. 'That's *miyamiya*', said Fred. 'It's what you get in Sickness Country if you bother the sacred sites.'

The mine came into view behind the chain-link fence on the right side of the road – pale spoil heaps crowded the foreground; behind them, the tanned cliffs of the Arnhem Land escarpment on the horizon and the looming presence of Djitbidjitbi, another *djang andjamun*, to the south. Djitbidjitbi, or Mount Brockman, is said to be the home of Dadbe, a king brown snake who is cousin to the Rainbow Serpent herself. If Dadbe is ever disturbed, say the Mirrar, she will unleash a flood of world-ending proportions. Despite this warning, Australians have dug uranium from the ground in the shadow of the holy mountain since the early 1980s. Here, in 1969, geologists discovered the richest uranium lode in the southern hemisphere. It was the most potent find since the 1915 discovery in the Congo of the famous Shinkolobwe mine, which produced the uranium used to level Hiroshima and Nagasaki.

I had never seen an open-cast mine before. I'd inquired about a guided visit but had no luck: tours of the mine were no longer on offer. Still, I knew that the road up to the gate was open and I wanted to see, if I could, this new manifestation of Sickness Country. Since Ranger opened in 1981, there have been more than two hundred spills and accidents that released hazardous material into the environment. In 2004, the operators discovered that miners had been showering in and drinking radioactive water four hundred times above the limit for human consumption. In 2010, millions of litres of uranium-contaminated water were released into the Kakadu wetlands. A burst leach tank in 2013 leaked one million litres of acidic uranium slurry. Although this time it did not breach the limits of the national park, a toxic mix of crushed ore and sulphuric acid covered part of the mine in a red crust three centimetres thick.

There were no diabolical scenes today, though – whatever I had imagined I'd find. It was quiet as I parked, and the heat filled my lungs as soon as I stepped from the car. Three birds of prey turned in lazy circles above the powder-blue spoil heaps. I took a few photos through the fence, although there wasn't much to see: I could glimpse the silent centre of operations in the distance, a gleaming conurbation of pylons, walkways and pipeworks, but the area immediately in front of me was scrubby and neglected, a dumping ground for industrial detritus. A few miners in orange-and-blue uniforms walked past after their shift, stowing their white hard hats in a nearby trailer, but no one bothered me. It seemed they were used to visitors poking about the perimeter. A kite dropped into the car park and began to sport back and forth over the fence, where I couldn't go.

The whole place seemed to be exhaling slowly. It was the end of the day, but also very nearly the end of the life of the mine. Rio Tinto's lease was up in 2020; only a few more years of operation remained before the mine would close for good. The plan is then to set about healing this particular Sickness Country. According to a report by the Australian government, the company will rehabilitate the 'disturbed sites' in and around the mine, restoring the abundance of wildlife and flora, and remove or somehow make secure around thirty million tonnes of highly radioactive mining waste.

Uranium is immensely old, older even than Earth itself – it's thought that all uranium on the planet originated in the furnaces of supernovas more than six billion years ago. It is also the heaviest natural element, so large it strains against the limits of itself. Like ball bearings shaken in a paper bag, its ninety-two protons perpetually threaten to tear the atom into pieces. The kiss of a single alien neutron can be enough to set off a chain reaction of astonishing power, as the isotope strews its particles in all directions with destructive force. Tom Zoellner, the author of a history of nuclear energy, compares the self-destructive frenzy of a uranium atom to a delusional man tearing off his clothes. Its heaviness makes uranium-235 vulnerable to splitting, but the source of its agitation is simply the urge for stability. To say uranium atoms decay is really to say they submit to a form of compulsive reorganization, producing new isotopes in a steadily stabilizing chain that ends in the mute fastness of lead-206. On the way, however, it releases a range of ionizing particles known as decay products. These particles, shed from the nuclei at 150,000 miles per second, ricochet through whatever living tissue they come into contact with, stripping atoms of their electrons and leading to mutation or malfunction.

Each decay product attacks the body as though it were possessed by a special desire for a particular living tissue. Radium-226 invades the teeth and bones, as well as mother's milk; radon-222 assails the lungs; caesium-137 plunders the muscles. Strontium-90 actually bonds to the structure of bone and gathers in the vascular tissues of plants.

I took a few more pictures, then got back into my car and turned out of the car park. Despite the warning signs, after a few hundred metres I paused on the road out to look through the fence again at the rim of one of the empty pits. Its sides looked emaciated, ringed by access roads like immense ribs. Thick, lime-coloured water pooled at its base. A tangerine excavator stood patiently on the far ridge with its arm curled in gentle repose. Beyond, in the distance, was the square, impassive face of *Djitbidjitbi*.

Perhaps, though, the mine didn't take too kindly to visitors after all. I had only stopped for a few minutes when a car pulled up beside me and I was asked, politely but firmly, to move on.

THE NEW SICKNESS COUNTRIES are our own creation. Their names flash and echo: Fukushima Daiichi, Windscale, Enewetak; Hanford, Mailuu-Suu, Lake Karachay; Mayak, Pripyat, Yucca Flat. Like the storm-cloud-bodied serpent of Wright's imagining, the mushroom clouds of the atomic age have heralded a remaking of the earth.

When the first atomic bomb was detonated at Alamogordo, New Mexico, on the morning of July 16, 1945, the blast vitrified the surrounding desert. Superheated sand thrown into the air liquefied instantly and then cooled as it fell in a storm of pale-green glass hail, filling the bomb crater with what looked like a lake of jade. Wide-scale thermonuclear-weapons testing began in 1952, reaching a peak in the early 1960s. Since the Trinity test at Alamogordo, more than sixteen hundred nuclear devices have exploded around the world – on average, roughly one detonation every ten days for forty-two years.

Perhaps the most infamous experiment took place at dawn on 1 March 1954, when the US military detonated 'Bravo', a fifteen-megaton bomb, over Bikini Atoll in the Pacific. The blast vaporized three small islands and left a crater a mile wide. The fallout was detected in rain that fell on Japan and in wind that blew across Australia. Marine life was so saturated by radiation that fish caught around Bikini would leave behind a spectral image if pressed onto photographic plates, the most highly radioactive parts of their bodies flaring as if lit by an inner explosion. On Rongelap, an atoll 140 kilometres away, islanders gathered on the beach on the morning of the Bravo test to watch the spectacle. Witnesses described a brilliant light on the horizon, like a second sun. Not long after, snow appeared to fall on the beach. They had heard about snow from the missionaries; that it would fall on the Marshall Islands seemed like a miracle. Children frolicked in the dusting of white particles that covered the sand, catching them on their tongues. Only later did the islanders discover that the snow was in fact a compound of vaporized coral and radioactive ash.

Exposure to the fallout caused blisters, hair loss, and radiation sickness in the Rongelap islanders, and they were evacuated to Kwajalein Atoll in 1954. Scandalously, they were returned to Rongelap in 1957, but Bravo's second sun continued to cast its baleful light. Long-term and intergenerational health problems began to occur. The islanders were afflicted with dozens of different forms of cancer, especially of the thyroid gland. Women experienced reproductive traumas, suffering miscarriages and, occasionally, nightmarish births.

In a testimony given to the International Court of Justice in 1995, Lijon Eknilang, a witness to the Bravo test, said that sometimes Rongelap women would give birth 'not to children as we like to think of them, but to things we could only describe as "octopuses", "apples", "turtles" and other things in our experience'. These 'jellyfish babies' – born without bones, and with skin so transparent their brains and beating hearts are visible – live for only a few hours, trembling on the cusp of unreality. The islanders were removed for a final time in 1985, and Rongelap was declared unsafe for human habitation for twenty-four thousand years.

In the Marshall Islands, making new Sickness Country also meant creating whole new landforms. In Enewetak, a thin, elliptical loop of forty coral islands, one island was entirely vaporized, leaving behind a two-kilometre-wide crater. In all, there were forty-three nuclear tests on Enewetak between 1948 and 1958. In the late 1970s the US government gathered up eighty-five thousand cubic metres of radioactive topsoil, including huge quantities of plutonium, and deposited it on Runit Island in the hundred-metre crater left by 'Cactus', the code name for one of the last bombs tested. They covered the crater with a domed concrete shell half a metre thick. The people who still live on Enewetak call it the Tomb.

Aerial photographs of the Tomb show another crater alongside it, almost exactly the same size, which has filled with seawater to form a new lagoon. Two halves of a figure eight, convex dome and concave lagoon mirror each other. The engineers neglected to line the bottom of the Tomb, which is now filling with seawater, its surface cracking like leather left out in the sun. It lies at sea level, and the water is creeping ever closer to its edges.

In all, five hundred atmospheric tests took place before the Test Ban Treaty in 1963. During peak testing, the quantity of radioactive carbon isotopes in the atmosphere doubled; although this has receded, traces of the twentieth-century 'carbon-14 bomb pulse' will remain detectable for the next fifty thousand years. Isotopes of plutonium-239 generated in nuclear reactions have a half-life of 24,100 years; Pu-239 is effectively nonexistent in nature, but residual amounts can now be found all over the world. Fallout from this carnival of atomic energy has left traces at both poles and on every continent, in lake sediments and ice cores, in tree rings and living tissues; it's so extensive and evenly distributed that many scientists believe it will be the most enduring signature of the Anthropocene.

Alongside the fallout of nuclear testing is the problem of waste left over from generating nuclear energy. Nearly all uranium ore – above 99 per cent – is U-238, a non-fissile isotope with an extremely slow rate of radioactive emissions. The half-life of U-238 is 4.5 billion years, roughly the same as the age of the planet. To be made into a viable fuel, the fissile quotient in natural uranium ore, U-235, with a half-life of 703.8 million years, must be recovered and concentrated. Milling and enrichment bring the negligible level of U-235 in natural ore bodies (around 0.7 per cent) up to between 3.5 and 5 per cent (for weapons-grade uranium, the U-235 component must be 95 per cent). But it's a highly profligate process. One tonne of natural uranium will yield 130 kilos of fuel and 870 kilos of mine tails, finely milled waste matter that still contains the majority of the radioactive material. Used fuel consists of approximately 1 per cent U-235 and 1 per cent plutonium; more than 95 per cent of depleted uranium isotopes are U-238. Some of this material can be recycled – although it is not fissile, U-238 is known as 'fertile' because it can 'capture' an additional neutron to become Pu-239. Reprocessing waste in this way can reduce the amount of depleted fuel by 80 per cent, but it remains dangerous to life for thousands of years.

Nuclear waste products present us with futures beyond our grasp, promising harm to bodies unborn and landscapes that have yet to emerge. They are our point of entry into what the writer Russell Hoban calls 'the moment under the moment': the depths that plunge beneath the surface of the everyday. Behind the shared sense of the real that allows us to work and live alongside each other is another reality, Hoban says, one that can be approached only in 'the flickering of seen and unseen actualities'.

On 26 April 1986, the flickering moment beneath the moment erupted into vivid life. The failure of reactor number four at the Chernobyl nuclear power plant in Ukraine released fifty million curies of radioactivity in a single explosive instant. A sense of unreality took hold. The needles of pine trees close to the blast turned red and refused to decay when they fell; fallout burned holes in the leaves of cherry trees. In her remarkable oral history *Chernobyl Prayer*, Svetlana Alexievich recounts the testimony of a woman who found glistening blue lumps of caesium in her vegetable plot.

But most dangerous of all was that the newly lethal environment mimicked everyday reality almost perfectly. In the immediate aftermath, life seemed to go on as usual – children played outdoors, bakeries sold bread in the open air – unheeding of the fact that the area was being drenched by an invisible downpour of radionuclides. *Do not eat the bread* became a warning passed between those who understood the seriousness of the disaster. By 5 May, the fallout had reached India and North America.

Two people died in the initial explosion, followed by twenty-nine first responders in the several days afterwards. These are the only deaths explicitly linked to the catastrophe, but it is thought that it will have caused upwards of forty thousand cases of cancer by 2065. Beyond this slowly unfolding disaster, though, another more inexorable moment broods beneath the present. Like the atolls of the Marshall Islands, the site will be uninhabitable for the next twenty thousand years, twice all recorded history. If we follow the same reach of time into the past, it takes us to before the origins of writing or indeed any recognizable language spoken today, to the very beginnings of metalworking. Joseph Masco calls this the 'nuclear uncanny', a torquing of our sense of time by the complementary dangers of 'the millisecond and the multimillennium'. Chernobyl is frozen in its explosive moment for the next twenty millennia; from a human perspective, it will always be 1.23 a.m. (UTC +3) on 26 April 1986, inside the vast concrete sarcophagus around the ruptured fourth reactor.

Our capacity to split the atom and mine the energy encased in the stuff of life itself can seem like humanity's apotheosis. Robert Oppenheimer is famously said to have borrowed from the *Bhagavad Gita* as he witnessed the first nuclear detonation over Alamogordo: 'Now I am become death, the destroyer of worlds.' But the story is apocryphal, and disguises the fact that our capture of the uranium atom's power has not made us gods, but turned us into supplicants to the new immortals we have fashioned. And like the gods of old, it seems inevitable that these novel deities – powerful, unchanging, and capable of invisible and soundless damage – will demand their own body of myths, ones that tap into the nuclear moment thrumming beneath the present.

In 2018 the Marshallese poet Kathy Jetñil-Kijiner set out on a four-day canoe journey to Runit to visit the Tomb. 'I'm coming to meet you', she says in 'Anointed', a video poem of the journey. 'What stories will I find?' Standing at the top of the concrete dome, she recounts the story of the trickster Letao, who received a gift from his mother, the Turtle goddess: a piece of shell with the power to transform him into whatever shape he wanted – a tree, a house, even another person. But Letao uses his gift to turn himself into kindling to make the world's first fire, and gives himself to a small boy who nearly burns his village to ash. 'While the boy cried,' Jetñil-Kijiner says from the apex of the cracked concrete shell on Runit, 'Letao laughed and laughed.'

'The fabric of our myths and folk-tales is in us from before birth', states Hoban, for whom 'the patterns of blue-green algae and the numinous wings of the Great Nebula in Orion and the runic scrawl of human chromosomes are stories'. So, too, are the ninety-two straining protons in an atom of uranium. Moreover, it is certain that these won't be stories of benign creation. Lake Karachay, in eastern Russia, is so polluted, it's said, that standing in it for an hour would be fatal. A thousand years or more from now, when memories of the nearby Soviet nuclear factory have evaporated, human invention will need to bend towards explaining the burning antipathy that lies beneath the lake's savage waters.

The nuclear immortals will jealously guard the toxic earth they preside over. Perhaps not only myths and stories but also propitiatory rites will be required to stay their hands, which would otherwise for ever fill whole landscapes with illness and death.

OF COURSE, it's possible that someone might want to claim the power of the most polluted sites for himself.

In Sophocles' *Oedipus Rex*, Oedipus begs Creon – the king of Thebes who succeeds Oedipus in his disgrace – never to condemn Thebes to house his corrupting body. Blind and frail, an outcast from civilized life, Oedipus carries within himself a dread certainty. 'No sickness can destroy me', he proclaims. 'Nothing can . . . I have been saved / for something great and terrible, something strange.' But his warning is not heeded.

Oedipus is the ultimate contaminated entity, placed beyond purification by the twin blemishes of patricide and incest. Sophocles calls him 'a man stained to the core of his existence'. Yet in this pollution there is power. In *Oedipus at Colonus*, the sequel to *Oedipus Rex* and Sophocles' final surviving play, the dying Oedipus arrives at a sacred grove on the outskirts of Athens in search of refuge. Creon seeks him out so that he can bury him at the Theban border, where his grave will confer unmatchable protection on the city. Yet Oedipus refuses, calling down furious curses on Thebes and its king that will see his 'vengeance / rooted deep in your soil for all time to come!' Instead, he gives his body over to the trust of Theseus, king of Athens, to be buried in secret so that it will be a perpetual defence for Athenians and a blight to the Thebans. As reward for hosting 'the power that age cannot destroy', Oedipus promises Theseus that his hidden remains will be 'the root of all your greatness, everlasting, ever-new'.

The death of Oedipus is not the only story from classical times to tell of a secret tomb that is the source of tremendous power. Herodotus describes how the Spartans are continually bested by their enemies, the Tegeans, until they inquire of the oracle at Delphi. The priestess promises that they will see victory only once they find the resting place of Orestes – Homer's 'far-famed son of Agamemnon', who killed his mother, Clytemnestra, and her lover, Aegisthus, to avenge his murdered father – and bring his bones back to Sparta. Their efforts come to nothing, until a further hint from the oracle leads one of the Spartans, Lichas, to a blacksmith's yard. According to the oracle's clue, Orestes has been interred in a place of smiting and counter-smiting. As Lichas watches him at work, the blacksmith recounts his astonishing discovery of an enormous coffin up to ten feet long, containing a body equal in length, as he was digging a well in his yard. After telling his fellow Spartans what he has learned, Lichas returns to the forge, claiming to have been exiled, and persuades the blacksmith to lease the yard to him. Having it finally to himself, he digs up the bones of Orestes and returns to Sparta in triumph. 'Ever since that day,' writes Herodotus, 'the Lacedaemonians in any trial of strength had by far the better of it.'

The story of Orestes and the Spartans may have its future equivalent. Alexievich describes how the first casualties of the Chernobyl explosion, including the plant operator who initiated the emergency shutdown, are buried in Moscow in zinc-lined coffins under one and a half metres of concrete layered with lead. That such extraordinary measures were taken to seal an individual in the ground will surely provoke wonder in anyone who might discover the graves far in the distant future. We assume that the elaborate care taken over ancient burial sites is an index of the power wielded by the individual interred, or the respect owed to them in life. Places like this provoke curiosity, and the interest of those who think they might find concealed within a treasure to enhance their own wealth or status. What we hide in the earth, fearful of 'the power that age cannot destroy', may not stay hidden.

INCREDIBLY, despite its deadly potential, no one seems to know exactly how much nuclear waste exists. In 2007, the International Atomic Energy Association attempted a global inventory, which estimated that 2.2 million tonnes of enriched uranium had been produced worldwide, along with as much as 220 million tonnes of radioactive mine tailings. Ranger alone has yielded between 3,300 and 5,500 tonnes of fuel per year since it opened; in just the ten years after the IAEA report, it produced over 35,000 tonnes of triuranium oxide (otherwise known as yellowcake), all of which was shipped to power stations around the world, including Fukushima and Torness, near Edinburgh, in sight of the beach I visit each spring with my students. Suffice to say, there's a lot of it, and the problem it poses isn't going to go away. There are 450 operative nuclear power stations around the world, and most of them store their fuel aboveground, vitrified inside steel casks. But these are only short-term solutions, and given the length of time that must elapse before the waste is no longer a threat to life, they are unconscionably vulnerable.

Safe long-term storage of spent nuclear fuel is a serious conundrum. International treaties forbid burying it beneath the sea or in the Antarctic ice sheet. Nor can we eject it into space, unless we can live comfortably with the risk of rocket failure spraying spent fuel into the atmosphere. To address this issue, engineers and scientists from a number of nuclear states have spent decades and billions of dollars developing sites deep in the earth that may operate as reliable repositories of even high-level nuclear waste. Their efforts have been beset with difficulties. In 2002 the US Congress approved a plan to develop a deep-earth storage facility for seventy thousand tonnes of spent fuel at Yucca Mountain, Nevada, a hundred miles from Las Vegas, but funding was withdrawn nine years later as projected costs rose to nearly a hundred billion dollars. In 1999, the Waste Isolation Pilot Plant, or WIPP, near Carlsbad, New Mexico, received its first deposit of transuranic waste – mostly contaminated protective clothing and laboratory equipment, as well as a solidified toxic slurry generated in the production of nuclear weapons. This was the first time artificially created radionuclides had been stored underground. The plan is to continue filling WIPP until 2030, then make it safe from intrusion for at least the next ten thousand years.

The area around Carlsbad was chosen because it lies on an ancient ream of salt – the Permian-era Salado formation – laid down when coral reefs sealed off parts of a shallow sea around 250 million years ago. The restricted water slowly evaporated, leaving behind a layer of crystalline salt that today is between two hundred and four hundred metres thick. The Salado makes a highly effective tomb because salt spreads: gradually, the walls of the chambers inside WIPP will creep inwards until the waste is completely engulfed. It also prevents the flow of groundwater and has a high degree of plasticity, protecting the contents against tectonic disturbances: if in the future the tomb is cracked by earthquakes, the creeping salt will heal itself.

Human intrusion, however, is another matter entirely, especially as a fifty-four-million-barrel reserve of fossil fuels lies buried just a kilometre from the site. Communicating the risk posed by WIPP to generations of people who will not be born for thousands of years is an unprecedented problem. The challenge was to devise a message that will remain active – legible, interpretable, and thus effective – for longer than there have been cities, or writing.

All words have a half-life. Typically, this can range from 750 to, in extremely rare cases, more than 10,000 years. Use wears down the value of some words and modifies others, but like radioactive elements, all are subject to the same inexorable decay. Palaeolinguists call this *semantic* or *phonetic erosion*, a force my students meet with when they first read the strange injunction to pay attention, 'hwaet' (which means 'listen up!'), that opens the Old English poem *Beowulf*. The example of the Yidindji shows that words can be preserved in place names, and some 'ultraconserved' words do persist, usually pronouns and numerals: in English, 'I', 'you', 'not', 'here', and 'how' are thought to be twenty thousand years old. But the remains are scant and insufficient for effective communication. One of the reports that examined the problem of communicating with the deep future anticipated that, ten thousand years from now, English – if it is still spoken – will retain only 12 per cent of basic words currently in circulation.

The problem briefly gave rise to what is perhaps the most remarkable branch of knowledge ever devised in the study of human language. It was known as nuclear semiotics, and its founder was an American semiotician, Thomas Sebeok. In 1984 Sebeok published a short paper with an astonishingly long imaginative reach, in which he proposed that the best way to protect a message against the erosive effects of deep time was to form what he called an atomic priesthood. Sebeok put his faith in continuity and the power of myth. Superstitions should be allowed to accumulate around sites like WIPP, he said, wreathing irradiated landscapes in an aura of illness and threat to discourage the curiosity of the uninitiated. Sebeok's atomic priests – 'a commission of knowledgeable physicists, experts in radiation sickness, anthropologists, linguists, psychologists, semioticians and whatever additional expertise may be called for, now and in the future' – would guard the truth of what lay buried, in the form of a specially designed annual ritual. To inoculate against decay, Sebeok devised a relay system for transmitting information, in which the message would be renewed on a three-generation cycle: designed to be understood, therefore, by the great-grandchildren of each current priesthood. Such a measured evolution would, he argued, keep pace with linguistic change.

As he faces his own entombment, Oedipus presents Theseus with a similar injunction. 'But these are great mysteries . . . / words must never rouse them from their depths', he warns:

> *When you reach the end of your own life,*
> *reveal them only to your eldest, dearest son,*
> *and then let him reveal them to his heir*
> *and so through the generations, on forever.*

Sebeok's relay, transforming as it migrated through time, would be supplemented by a 'metamessage', with an injunction to future generations to maintain and update it every five hundred years, coupled with veiled threats that 'to ignore the mandate would be tantamount to inviting some sort of supernatural retribution'.

In the early 1990s, two teams were commissioned to devise a potential marker scheme for WIPP. One would focus on engineering the landscape to convey a warning, the other on a mix of written and pictorial messages. The final report proposed making the entire site a kind of full-body communication system, engaging both the mind and all the senses in an experience of dread. The barriers would be purely symbolic; effectively, the markers would depend on evoking a particular emotional response in people whose cultural norms may be as distant from ours as those of ancient Egypt. Consequently, the schemes proposed by each team imagined a series of truly fantastic worlds. Landscapes of brutal thorns, lightning-bolt-shaped earthworks and immense spike fields, like the work of a race of demented giants. One design centred on a large pool of black-dyed concrete, a void that would absorb the desert heat and radiate it back. Another proposed to bury noxious and even radioactive materials in durable glass capsules that would rupture if disturbed.

The final design, announced in 2004, involved five levels of warning messages, rising in complexity, and a mix of monoliths, buried clues, and archives. A perimeter of twenty-five-foot granite markers will patrol the limit of the controlled area, with further markers outlining the repository footprint. Each one will be engraved with a message, in seven languages (Mandarin, English, Spanish, French, Russian, Arabic and Navajo), prohibiting digging or drilling in the area for ten thousand years; their lower portions, buried seventeen feet down, will also be engraved with the same injunction. The message will be accompanied by faces of disgust and repulsion, modeled on Edvard Munch's *The Scream*. Nine-inch ceramic disks, buried at random between two and six feet down within the repository footprint, will also carry an abbreviated warning and Munch's icons of existential horror. A thirty-foot earthen berm, studded with magnets and radar reflectors to signal an anomaly, will enclose the inner ring of granite markers.

Two storage rooms will be buried twenty feet below the surface, one within the berm and one outside, with a single conical opening. Anyone who chooses to ignore the messages aboveground and break into these rooms will find the granite walls inscribed with more detailed warnings about the dangers of pressing further; or they might encounter the 'hot cell', an archaeological remnant with low-level irradiation where transported transuranic waste was transferred from 'road' casks to storage casks before WIPP was sealed.

At the very heart of the site, an information centre will house a spatial perspective of the repository and a geological cross section; a copy of the periodic table; star maps of Vega, Arcturus, Sirius, and Canopus to indicate when the site had been built; and a world map showing the location of other repositories. A complete archive of WIPP and its contents will be housed off-site, presumably to be maintained by a version of Sebeok's atomic priesthood.

Should they be built, the WIPP markers will describe a landscape ready-made to be mythologized, broadcasting their fearful message long after their creators are dust. But the nature of those myths – whether they declare our civility or our savagery – is beyond the control of anyone living. There seems to be something perversely self-aggrandizing about them: not just a warning, but a testament to the heights of destruction our civilization could achieve. Perhaps, then, their real counsel is for us. Like the chorus in a Greek tragedy, the dozens of *Scream* faces scattered around the WIPP site will sorrow over the Sickness Countries we've made.

Sophocles wrote about the tussle over Oedipus' remains just before his death in 406 BCE, but by the time *Oedipus at Colonus* was performed, five years later, Athens had been devastated by war. It would never recover its former glories. Athenians in the audience at Sophocles' last play, listening to Oedipus promise Theseus eternal protection in return for safeguarding his buried corpse, would have recognized the hubris in assuming the future could be guaranteed against calamity, and must have felt they were watching a lament to their city delivered from beyond the grave.

LIGHT RAIN MISTED the fields and small woods. There had been a heat wave, and though cooler now, the air was still close as I crossed the island of Olkiluoto, on my way to Onkalo.

Olkiluoto is an island of roughly ten square kilometres on Finland's Bothnian coast, blanketed in spruce, black alder and pine trees, which, like the deep forests of this tree-loving nation, stretch unbroken across mile after flat mile. The island is the location of one of Finland's two nuclear power stations; it is also home to what will be the world's first deep geological repository for spent nuclear fuel. The method is, on the face of it, disarmingly simple: dig down into the ancient bedrock; bury the waste in specially designed copper canisters; backfill the hole; and retreat, without leaving a single trace aboveground.

Unlike WIPP, conceived deliberately as a place that will foster new stories of Sickness Country to discourage potential invaders, Onkalo, the Finnish repository, is meant to be forgotten.

Onkalo will store all Finland's nuclear waste for the next 120 years, 450 metres down in the bedrock. Already, the Finns have completed 5 kilometres of tunnels large enough for trucks to pass through. Finland's geology is particularly suited to the long-term storage of hazardous material. It is situated on an immense granite laccolith, the Precambrian Fennoscandian Shield, consisting of igneous rocks laid down nearly two billion years ago. Although now buried beneath a layer of till deposited after the last ice age, Finland's bedrocks are in fact the roots of long-gone mountains. They form one of the most geologically stable and homogeneous areas on the planet, far from plate margins and riddled by a fracture system of fine veins, caused by the weight of ice sheets, that provides a natural plasticity, further cushioning against stress. Unlike WIPP, Onkalo has no valuable resources nearby. Groundwater flow is extremely limited – water below three hundred metres is likely to have been underground for most of the past two billion years. The Fennoscandian Shield has scarcely changed since the time when the only life on Earth was bacterial. Perhaps the Finns have no need of new myths, given that their nation stands on rock of such mythic hardness.

I had come to Onkalo not just because it was an opportunity to walk through the tunnels that would keep spent nuclear fuel in isolation for tens of thousands of years, but also because I was intrigued by the plan to leave the location unmarked. This seemed like either an extraordinary carelessness or an astute reading of the hubris involved in trying to communicate across deep time.

I had arranged to be part of a tour for international press and been instructed to convene at the power plant's visitor centre. The road to Olkiluoto passed through a flat, rural landscape, sparsely populated by just a few farms and isolated wooden buildings. A sign forbidding photography at the gated entrance to the plant was the only indication that something out of the ordinary was here.

I was early, the first to arrive. The attendant at reception took my name and suggested I fill the time before the rest of the press party arrived by looking around the exhibition. It told the story of nuclear energy, from mining the ore to harvesting the energy to final disposal, with all the clarity and simplicity of a child's fairy tale. Everything was clean and controlled. The centre was still bundled in its early-morning hush, but from the cafeteria I could hear the faint techno pulse of a radio playing 2 Unlimited's 'No Limit'.

At the end of the exhibit, an immense copper canister gleamed next to the icy sheen of what looked like a solid cylinder of iron with a grid of square holes punched through it. Both were four and a half metres long, and the copper canister was large enough for me to crawl inside comfortably. Laid prone at its mouth was the lid, a huge coin of gleaming copper. This is how Finland will protect its waste from prying eyes. The spent fuel will be cast in fingernail-size ceramic pellets and arranged in rods, which will be inserted in the square holes of the iron cylinder insert, then sheathed in the copper canister and sealed with its shining, bright lid. The canister will be borne to a depth of 450 metres inside Onkalo and deposited upright in an 8-metre hole drilled into the bedrock. The hole will be filled with a clay buffer to prevent movement and erosion by groundwater, and a concrete cap will be lowered into place. In all, there will be 5,400 deposition holes located in 137 short tunnels inside Onkalo. When it has received its full deposition, each tunnel will also be filled with bentonite clay and sealed with concrete, and then the whole repository will be backfilled to the surface and the entrance covered over. After that, it will be left for the forest to reclaim.

Finnish engineers scoured the world for analogues that would prove the durability of their design and materials. A bronze cannon that sank with the Swedish warship *Kronan* during the Battle of Öland in 1676 and landed muzzle-down in water-saturated sediment, demonstrated the resilience of the copper canister, having experienced less than 4 per cent corrosion in four hundred years. Plates of elemental (naturally occurring) copper found in a mudstone formation in Littleham, Devon, were largely unchanged after 170 million years. Intact Roman nails dropped in a pit in Inchtuthil, in Scotland, two thousand years ago, testified to the reliability of the iron insert. Two-million-year-old preserved sequoia trees found at Dunarobba, in Italy, buried upright in lacustrine clay by an ancient flash flood, and the 2,100-year-old corpse of a Chinese woman, found naturally preserved by a thick layer of clay around her coffin, her organs intact and joints still movable, confirmed the effectiveness of the bentonite clay as a barrier. Everything seemed methodical and soberly tested, in contrast to the air of fantasy that seemed to hang around the plans for WIPP.

But perhaps the Finns could have looked at another analogy closer to home. The evening before I visited Onkalo, I had gone down to the shore of the Bothnian Sea and, in the coppery light, read from the *Kalevala*, Finland's national epic poem. The tales and songs that make up the *Kalevala* were collected in the nineteenth century by a physician called Elias Lönnrot, mostly on foot. Each spring, he would set out to walk across rural Finland, discarding his shoes and coating his feet in tar and in this way covering over thirteen thousand miles in pursuit of the old stories, which he gathered up into a single epic narrative, modelled on *The Odyssey*. One of the strangest tales in the *Kalevala* is of the forging of the Sampo.

It begins when, to repay a debt he owes to Louhi, mistress of the dark Northland, the old trickster Väinämöinen deceives the blacksmith Ilmarinen. Väinämöinen convinces Ilmarinen that in the far north lives a young woman of surpassing beauty who has refused every man who has approached her. But she will yield to the one who can forge the fabulous Sampo, a powerful and mysterious object also known as the 'bright-lid', which only a craftsman as skilled as Ilmarinen can beat out. Then cunning Väinämöinen sings the wind into a strong gale, which carries Ilmarinen up and draws him far north across land and water.

Louhi grins a gap-toothed grin when Ilmarinen alights, declaring that he has arrived to forge a new Sampo, and chides her daughter to dress in her finest clothes and put red on her checks: 'Ilmarinen, the everlasting craftsman has come,' she exclaims, 'to make the Sampo, brighten the bright-lid!'

Besotted by the beauty of Louhi's daughter, Ilmarinen sets about forging the Sampo from a swan's quill, the milk of a barren cow, a small barley grain, and a ewe's summer fleece. He finds a spot for his forge, builds his bellows, and lights a seething fire. Here he pushes the raw materials into the flames and charges the bellows for three days and three nights. On the first day, peering beneath the forge to see what's what, Ilmarinen finds a crossbow of gold, which he snaps in displeasure. On the second day, he withdraws a red boat, its prow golden-bright, and breaks it to pieces. On the third day, there comes a heifer with golden horns; on the fourth, a golden ploughshare. None please Ilmarinen, who has built an immense flame in his forge. Finally he peers beneath the forge and sees the bright-lid beginning to form. He hammers the Sampo into shape with infinite skill. When he is done, there is a corn mill on one side, a mill for salt on a second and a money mill on a third.

'And then the new Sampo ground / and the bright-lid rocked', records the *Kalevala*, grinding one binful to be eaten, one to be sold, and a third to be stored away.

Ilmarinen delivers his prize to Louhi, who grins again and takes the Sampo deep inside the hard hills of Northland, within a slope of copper, where she hides it behind nine locks and secures it with powerful, binding roots to a depth of nine fathoms.

There is no account anywhere in the *Kalevala* of precisely what the Sampo looks like or what exactly it achieves, although it is clearly highly desired and thought of as the source of limitless wealth. One translator I read noted that 'the word stands somewhere between the glittering artifice of the Greek *daidalos* and the numinous radiance of the Welsh *gwyn* – where no English word is to be found.' The mysterious bright-lid lies gleaming at the heart of Finland's greatest epic poem.

The others began to arrive: a journalist from France and another from Italy, an Italian photographer, a Belgian architect, and Pasi, our Finnish guide. Trim and tidy, with a neat silver beard, Pasi was the communications manager for Posiva, the company responsible for building Onkalo. Over coffee, he explained that the Finnish approach to Onkalo was based on openness and trust. 'The less you know, the more you fear', he said with a shrug. Finland's used fuel will be stored above-ground and cooled for forty years, significantly reducing its radioactivity, before deposition in Onkalo. After five hundred years, he promised, the radiation dose from the spent fuel will be equivalent to the annual natural dose for one person; after ten thousand, it will be no worse than receiving an X-ray.

'I get asked this all the time', said Pasi. 'How can you put a time bomb in the ground? But after three or four hundred years it's no more dangerous than natural uranium – although you shouldn't eat it.' But I had read that Posiva had set a reference period of 250,000 years, to account for at least one future glacial cycle.

So why make it safe for such a long time? I asked. Why so much emphasis on ten thousand years? 'To feel safe', he replied. *Just* for the feeling? Pasi nodded.

Onkalo is scheduled to open in 2020 and will receive spent fuel until sometime in the middle of the next century, before the final seal is set in place. Pasi suggested that, given the long life span of the repository, it would be foolish to try to mark it. 'Eventually,' he said, 'another ice age will wipe out everything – maybe not for one hundred thousand years, but afterwards there will be no Paris, no London; nothing above-ground would survive the ice. If people returned here when the ice retreated, any marker left by us would be gone.' Even if it is delayed by climate change, a new ice age could smother all northern Europe in several kilometres of ice, which would press down on the bedrock. The design of Onkalo apparently takes this into account, but nothing would protect whatever was left at the surface. At different times it may be flooded beneath a shallow sea or exposed to the elements. The region around Olkiluoto continues to rebound following the retreat of the last ice sheets, at a rate of around six millimetres per year, and is expected to continue rising imperceptibly, but inexorably, for the next several thousand years. Posiva predicts that, ten thousand years from now, Onkalo will lie as many as fifteen kilometres inland.

Faced with such uncertainty, perhaps hiding the waste and trusting that people in the future will not be too curious is the most pragmatic solution. Trust certainly seemed to be the foundation of the Finnish approach, which not only seeks to build confidence in contemporary Finns but also puts its faith in the ingenuity and fidelity of future generations. Constructing Onkalo is a multi-generation responsibility, unlike any engineering project ever undertaken before. It's possible that the great-grandparents of the engineers who finally close it have yet to be born. Sealing Onkalo would be like bringing closure, today, to an endeavour begun at the end of the nineteenth century, in a world before air travel, the internet, or antibiotics, when even Finland itself – which gained independence from Russia in 1917 – did not yet exist as a nation.

We were joined by Anne Kontula, one of the hydrogeologists working on the excavation of Onkalo. I asked her how she felt about the prospect of communicating a warning to future generations. 'I think it's unrealistic', she said definitely. 'Think about how much can happen in one million years – that's not much in geological terms, but immense for humans'. I repeated Pasi's point about expecting another ice age. She nodded. 'Even with climate change, the planetary system of ice ages will continue. After five hundred thousand years carbon dioxide levels will be back to pre-industrial levels. But we trust in our bedrock.'

I showed Anne the final design to mark WIPP, including the rendering of Munch's 'scream' face. It was the first time she had seen it. She was polite, but evidently sceptical. The architect told us he had submitted a design that would involve building a spiral of ten thousand stones over the entrance to WIPP, like Robert Smithson's *Spiral Jetty*, each stone imprinted with a star map. Every year, a designated person would remove one stone until the entrance was exposed.

'We have another hundred years to decide what we do', said Anne. But you must have a preference, I said. Her reply was firm. 'I do – not to mark it.'

I rose from the table to get more tea. I found Pasi at the urn and asked him if he thought people in the future would tell stories about Onkalo after it was buried, like a new *Kalevala*: 'Good question!' he said. 'Some people do want to mystify it.' He told me the name Onkalo was first used informally among research scientists, but after a while it stuck. *Onkalo*, he said, meant a hole in which animals lived. 'Creatures in the forest live in *onkalo*, foxes and so on.' Like a burrow? I asked. 'Well, not quite', he said. 'It's a place you don't want to put your hand; something might bite you.'

WE BOARDED A MINIBUS and set off towards the repositories. There were two: one for low- and intermediate-level waste, and another, deeper one for high-level waste – Onkalo. 'This is old-growth forest', Pasi explained as we drove across the site. 'You can see white-tailed deer and moose, and sea eagles along the coast'. Two wolves had apparently recently been seen in Pori, fifty kilometres away. 'This is *Kalevala* landscape!' he said with glee.

As we passed through the nuclear plant itself, Pasi told us

about the plant's most unusual by-product. Some of the water heated to drive the immense turbines was pumped underground, so the soil doesn't freeze even during the harsh winters. Someone had taken this as an opportunity to cultivate grapes and ferment wine. 'We call it Château Onkalo', said Pasi. 'But', he added, as if the disappointment still lingered, 'it's not very good.'

Our first stop was at the facility for burying low- and intermediate-level waste, similar to the irradiated clothes and implements stored in WIPP. It could have been anything, a garage for site vehicles or an electrical substation: just a squat rectangle set within a hollow of earthworks that, I later learned, were built up from material excavated from Onkalo. Beyond the entrance, however, the road continued, sweeping down into the rock.

It was cool inside, and bright with strip lights. The exposed granite walls were streaked with white quartz, gleaming in the artificial light after so many aeons in the darkness. 'These rocks are 1.9 billion years old', exclaimed Pasi. I noticed deep parallel lines scored along the walls, each one ending at a shallow concavity – punctuation marks where explosives had been used to dig out the tunnel. I thought again of the two-hundred-thousand-year-old stone axe I'd held in the National Museum storehouse in Edinburgh, with its rippling bulbs of percussion, and the wave of connection across deep time that had passed through me as I grasped it. With each step we walked down through millennia.

At a farther gate we collected hard hats and took a lift down to where the waste was kept. As we stepped from the lift, the first thing I saw startled me: life, sixty metres below the surface. The wall by the lift was vivid with wet green moss, thriving in the electrical light.

Pasi led us into the cavernous storage room. It was the size of an aircraft hangar, with what looked like a raised dais at either end, each large enough for a game of football. The whitewashed walls shone, and the smell of fresh paint filled my nose. 'Under here is where we keep the waste', he said, pointing to each dais. 'Low-level in one, intermediate in the other. That smell is the radioactive clothing decaying.'

Before we left the chamber, each of us submitted to a test to check how much radiation we had absorbed. My reading was 2.2 millisieverts (mSv), about the same amount I'd absorb from background radiation in my own home over the course of a year. 'It's so safe we bring parties of schoolchildren down here', said Pasi. As we emerged back on the surface, I mimed relief at breathing fresh air again. We boarded the minibus. The next stop would be Onkalo itself.

At the gate leading to the repository, we were met by Jari, who would be our guide in the dark. His job was to blast away the rock. 'Do you enjoy it?' someone asked. 'Sometimes it's very nice!' he said, grinning. Again, we collected our safety gear, and Jari led us to another, much smaller van.

As I put my hand on the minivan's frame to climb into the back, Jari – who hadn't seen me – pulled the driver's door shut with a bang, missing me by millimetres. I yanked my hand away. 'Oh no!' he said, but I wiggled my fingers and grinned to show they were okay.

We crammed into the minivan, and I looked down at my

left hand: I had travelled hundreds of miles to be here, and but for a hair's breadth this would have been where my journey ended.

But we drove on. The road into Onkalo seemed to pour us into the rock, curving round like a giant tongue. It was a bit like being swallowed. Instantly, daylight vanished and we were bathed in a coppery sodium glow. We plunged downwards.

After five minutes or so, at a depth of about two hundred metres, we stopped so the photographer could take some pictures. Outside the van, the air was musty. The road sloped away into darkness in both directions. There were signs mounted at regular intervals along the walls, indicating distance or the location of fire points and emergency access, but these hints of human order did nothing to diminish the sense we were standing somewhere deeply primeval, both ancient and half-formed. The signs marking the emergency exits showed a blocky human figure running towards a dark doorway. There were more scores along the walls, as if giant fingers had been drawn through wet clay. Thin runnels of water dripped down, but there was no sign of life here as there had been in the other repository. Water seeps so slowly through the tight cracks of the Fennoscandian Shield, it may have been hundreds of thousands of years since it last fell as rain.

I asked Jari how much rock had been excavated to make the five kilometres of tunnels that had been built so far. He didn't know the total but guessed it was around one hundred thousand tonnes in the past year alone.

We progressed down to 450 metres and stopped at a cavernous junction at the very base of the repository. There were tunnels branching off in several directions, and immense, insectoid excavating machines. Occasionally, we had to step aside as trucks barrelled past. The air was appreciably warmer. Pasi told us that this is where they planned to build facilities for the workers involved in construction and disposal, including showers and even a cafeteria. A superfast lift will connect the repository to the surface. For a time Onkalo will function as a kind of subterranean village, filled with the sound of work and the voices of the people excavating it. Generations of working lives will pass through here, and when the final tunnel is sealed, all that will remain to tell about the village in the rock will be the memories borne by the last cadre of engineers.

While we explored the junction, I noticed that the walls at this depth seemed different from those of the other repository. To guard against rock falls, every excavated surface, including the ceiling, had been covered in a wire mesh and then sprayed with concrete. As I peered closer, I could see tiny metal fibres poking like hairs from a lumpy surface that looked a bit like porridge. I tried to imagine the weight of the rock above our heads, and the weight of years: the astonishing patience of the repository, tolerating our brief intrusion, waiting to be closed again.

Our last stop was to visit the research tunnels, where engineers are perfecting the method for final disposal. There were three: one was busy with engineers digging a borehole; another, where they had been testing how to close each disposal tunnel, was blocked by a concrete seal; but the last, in Goldilocks fashion, was open. It was much narrower than the access tunnels we had passed through, blind at one end and coldly illuminated by arc lights. The walls were draped untidily with cables, and a generator and spools of more cables had been left at the back. Set into the ground, evenly spaced, were three enormous circular concrete lids, each perhaps one and a half metres across with a small square hatch in the centre. Jari lifted the hatch in the first and shone his torch down into the cylindrical borehole, eight metres deep.

To my surprise, there was a flash of lime; it was a pool of water the same shade of green as I had glimpsed at the base of the open pit at Ranger. Jari said they dyed the water sometimes, to track where it flowed.

Despite the clutter of construction materials, the only word I could think of to describe the tunnel was *holy*. This is where the final disposal of spent fuel rods will take place, a sanctuary set aside for the deep future. It was, unexpectedly, profoundly moving. The disposal tunnels and access roads will be backfilled to the surface, but if anyone in ten thousand years' time was tenacious enough to dig their way down here (and it would surely be much easier to remove the clay and backfilled materials than to excavate the original bedrock), this is what they would see. I felt jolted forwards in time, as if my own feelings suddenly mingled with the future visitor's rush of excitement. As the concrete seals were lifted and the gleam of the copper canisters caught the light again after thousands of years of being buried upright in the dark, would they feel horror, exhilaration or reverence? I could imagine their gasps echoing in the small tunnel as they raised the concrete seals to reveal the shining, bright lid. Would the Sampo still be remembered here – might they even think they had discovered the source of the myth?

I asked Pasi if anything else would be left in the tunnels before they were backfilled. 'They'll probably take anything of value', he replied. But it looked to me that much of what would be left down here would speak of the people who laid this vault in the earth. Not just the disposal chambers and their thousands of canisters, but the mesh-and-concrete-coated walls, the deep score marks, the pipes that will carry hot water to the engineers' showers. I was startled to realize that of all the many roads I'd travelled since I had walked across the new bridge over the Firth of Forth, this one I was standing on, hundreds of metres below ground, was the one most likely to survive as a future fossil. When nothing of the road network remains aboveground, the forty-two kilometres of roads through Onkalo will still be here, backfilled but intact, with their smooth surfaces and arcane signs, like the emergency exit markers with their running figures. I wondered whether a future intruder would read these as a warning to flee from unseen dangers or an encouragement – an invitation to hurry on through the door into the dark. There will be other signs too. The site around Onkalo is peppered with drill holes from studies of the bedrock.

Posiva's report on final disposal predicts what will be left of Onkalo beyond one million years. Plate tectonics are unlikely to have much effect. Serial glaciations may erode the surface-level plugs and some of the backfilled tunnels; but equally, over this timeframe, sedimentation may push them deeper underground. After tens of millions of years, the repository will perhaps have risen to the surface, wearing away the access tunnels and exposing some disposal chambers to the air. If this happens, some deposited material is likely to be dispersed into the environment, but by this time it will be no more deadly than naturally occurring uranium. Over the very long term, it may even come to resemble the uranium bodies it was drawn from. The copper canisters may have been partially reduced to copper sulphides, but because of their low solubility they ought to be largely intact even at this late stage, raising the possibility that still-gleaming remnants of the final deposition may be excavated four hundred thousand generations after it was sealed underground.

Jari beckoned us farther down the access tunnel, where a new branch had been begun. It was totally dark, but as he shone a light on the walls they suddenly erupted with life. Our waving flashlights revealed a tangled undergrowth of green lines, spray-painted by geologists tracing the cracks and fault lines in the bedrock. Each one was numbered, and the numbers and red dots looked like flowers and buds – like vines living in the deepest rock, clinging to the dark walls of Onkalo for all time.

ILMARINEN'S STORY does not end with the forging of the Sampo, which does not stay buried nine fathoms beneath a hillside in a vein of copper for very long.

After he presents the Sampo to Louhi, Ilmarinen presents himself to her daughter, who refuses to contemplate either his hand or leaving the dark Northland. Dejected, Ilmarinen returns home, where cunning old Väinämöinen hears him muttering in his distress of the lost bright-lid. Seized with desire for the marvellous Sampo, Väinämöinen sets off for the north with

Ilmarinen and a fair youth called Farmind to steal Ilmarinen's fabulous creation.

Väinämöinen lulls Louhi's warriors to sleep with his enchanted song, Ilmarinen smears her nine locks with butter so that they do not squeak on their hinges, and Farmind goes in to retrieve the Sampo. No matter how hard he pulls, though, it will not move, so Väinämöinen takes an ox from a nearby field and ploughs through the roots holding the Sampo in place. The three stow it on their ship and launch out onto the waters to make their escape.

The song of the triumphant thieves carries across the waters, startling a crane, which takes off shrieking across the Northland and rouses Louhi from a deep sleep. She grieves over her loss and gives chase, and a great battle ensues. Running on foot over the tops of the waves, Louhi assails Väinämöinen's boat; changed into the shape of an eagle, she drags her talons along its sides; perching on the masthead, she nearly capsizes it, until Väinämöinen addresses her in desperation and asks if she would share the Sampo. In her fury, Louhi refuses, and in the struggle that follows the Sampo falls overboard. The three thieves watch the dark water close over the bright-lid, where powerful waves break it in pieces.

With a shout of despair, Louhi flaps off to the dark Northland with the bright-lid's handle grasped in her clawed foot. Väinämöinen and his companions limp home, to find that what seems like their defeat is in fact a victory. The shore where they make landfall gleams with scattered pieces of the shattered Sampo. Väinämöinen gathers them up and builds an iron fence and a 'stronghold of stone' to guard them. The tale ends as old Väinämöinen sets up his prayer to the future, that neither sun nor moon will shine ill on his descendants, and that no enemy will steal the wealth guaranteed by the fragments of the bright-lid, rescued from nine fathoms deep inside the Northland hillside.

IN *CHERNOBYL PRAYER*, Alexievich recounts the testimony of one of the soldiers sent to assist with the 'clean-up' in the days immediately following the disaster. Rather than protective clothing and specialist equipment, the soldiers were handed shovels and buckets and expected to dig wearing only their uniforms. Their job was to bury everything they saw: trees, vegetation, even the top layer of soil. 'We buried earth in the earth', he recalls. 'Along with the beetles, spiders, and maggots, that whole separate nation. We buried a world.'

For Russell Hoban, the icon of the persistent impulse to tell stories was the severed head of Orpheus, raging and lamenting his terrible fate into eternity. We have also buried something like this severed head singing its fearful, ageless song. Burying some of the most dangerous materials ever devised by humans so far underground is a gesture of hope. We hide nuclear waste with such care because we do not want it to harm our descendants or make the land unfit to live on, but we also, I suspect, do so to protect our memories against their censure. Places like Onkalo and WIPP, great holes deep in the earth filled with long-lived radioactive by-products, hold more than the residue of our atomic adventures. We are buried there as well – or at least an image of ourselves that we would like to be forgotten. We bury the idea that we are an unprecedented threat to the future. Like Oedipus, we hope instead that what remains of us will find rest far from the city, that the dust will cover us and our sins fall out of memory. And somewhere deep within is the hope that the world we made – of multiplying Sickness Countries and death that visits unseen, and the flickering moment under the moment – can be safely buried and left behind as well.

SEVEN

WHERE THERE SHOULD BE SOMETHING, THERE IS NOTHING

At around 10.00 p.m. on 28 June 1927, Virginia Woolf boarded a night train at King's Cross station bound for the north of England. It was a Tuesday. Her companions were her husband, Leonard Woolf; her nephew, Quentin Bell; Vita Sackville-West (the model for the fluid-gendered hero/ine in Woolf's *Orlando*, which was published the following year); and Sackville-West's husband, Harold Nicolson. Woolf wore a fur-lined coat and smoked a cigar. The train was uncomfortably full, but Nicolson and Sackville-West nonetheless managed to fall asleep, his head curled on her lap. As they passed a level crossing, they saw a great snaking line of waiting vehicles, their lights burning patiently in the dark. At 3.00 a.m. the party ate their sandwiches. Half an hour later, the train arrived at Richmond in Yorkshire. Woolf and her companions collected their things and boarded an omnibus for Barden Fell.

When they arrived, they found the fell as crowded as their

train had been. More cars, some with people camped out alongside them holding impromptu picnics on spread mackintoshes; farmers with their families, dark and neat in their Sunday best. They were all gathered for one purpose: to witness, at dawn, the first total solar eclipse in Britain for two hundred years.

It was a cold, pale morning. The crowd was subdued, bound in silence as if awaiting the arrival of a marvel. The scene seemed to Woolf like a druidic memory, the ridge lined with expectant figures like Easter Island statues. 'Our senses had oriented themselves differently', she later recalled. The ordinary and workaday fell away: 'We were related to the whole world.' The ground was boggy and the sky thick with clouds. Their feet were wet. They stamped them to keep warm and worried that the blooming clouds would block their view.

Suddenly, through a gap, the sun appeared and began to race through the clotted sky, flashing brilliantly through the muffling dawn.

The cold air stiffened. Blues deepened to purples; faces took on a greenish underwater hue. Colour began to go out of the world. 'This is the shadow', the friends whispered to one another; 'this is the shadow.' Darkness loomed over the moor, Woolf said, 'like the heeling over of a boat', slow but inexorable until the crisis is reached, and capsize is inevitable. Then, abruptly, all light and colour was extinguished.

The eclipse was brief, no more than twenty-four seconds, and the light's return seemed like the world's remaking; and yet, Woolf wrote later in her diary, 'we had seen the world dead.'

Woolf recounted the eclipse in 'The Sun and the Fish', an essay on the way memories become attached to one another when they are retrieved from the mind's deep pool. The recollection of Barden Fell is joined by a visit on a stiflingly hot day to an aquarium in Kensington Gardens. Outside, London's summer thrums oppressively; but in their tanks, the blue and silver fish move with total equilibrium through sunlit waters. Luminous, almost transparent, they seem to Woolf like images of a grand design, of form aligned completely with being.

She drifts into a longing reverie; then, as abruptly as the shadow blanking the face of the sun, the eye of memory shuts, leaving, like a retinal imprint, only the image of 'a dead world and an immortal fish'.

'HOW CAN I EXPRESS THE DARKNESS?' wrote Woolf in her diary. 'It was a sudden plunge, when one did not expect it.'

The history of the planet is punctuated by extinctions. Many put only a small dent in global biodiversity, but there have been five mass extinction events when biodiversity crashed by at least 75 per cent: the Ordovician-Silurian; the Late Devonian; the Permian; the Triassic-Jurassic; and the Cretaceous-Tertiary, which saw the end of the dinosaurs and the rise of mammals. Most dramatic was the Permian, 225 million years ago, known as the Great Dying, when a rapidly warmed atmosphere and acidified oceans (due to a sudden and massive increase in volcanic activity) killed off 96 per cent of species, including all coral reefs. From a planetary perspective, extinctions are typically fairly sudden events. This is self-evident in the case of an asteroid strike, but volcanic activity or other climate factors can produce rapid swings in global temperatures, the composition of the atmosphere and ocean chemistry, which have a dramatic impact on life. It's estimated that the Great Dying took only around sixty thousand years; in the context of Earth's four and a half billion years, no more time than a twenty-four-second eclipse would be in the span of a day.

Today, rates of species extinction are rising – although estimates vary from one hundred to one thousand times the typical background level, no one disputes the direction of travel or the cause. Hunting has run many of the most charismatic species into the ground. Habitats are shrinking and resources for life are becoming scarcer under the pressures of human development; climate change is altering the terms that bind an animal to its environment. In 2019 the Intergovernmental Science-Policy Platform on Biodiversity and Ecosystem Services warned that up to a million species could be threatened with extinction – 12.5 per cent of all animals, plants and insects thought to exist today. But only a quarter of those at risk have been studied in detail. The majority constitute a vast blank, likely to be gone before their presence can make a mark on human speech. We will never really know what was lost.

Humans evolved in a world of teeming biodiversity, but we have tipped the balance to an extraordinary degree. The biomass of all wild animals is now less than a tenth of that of humans alone; if livestock and pets are included, then we and the animals we like to eat or live with account for 97 per cent of the total biomass of land mammals. The fossil record will reflect this homogeneity too, as the same bones of just a handful of domesticated creatures turn up again and again on every continent except Antarctica.

In every part of the world, the shadows are racing onwards;

wherever there once was colour, life is collapsing into darkness and silence. Some predict we are on the brink of a sixth mass extinction event, as each day up to two hundred species fall under permanent shadow. Unless we arrest that trajectory, we will certainly witness extinction on a par with the late Quaternary period, between thirty thousand and fifty thousand years ago, when around half of large mammals died out. The result will be a gap in the future fossil record, an evolutionary hiatus that will take millions of years to restore.

Woolf's account sent me back to Annie Dillard's essay 'Total Eclipse', which describes a solar eclipse in Washington State on February 26, 1979. For Dillard, the experience was like falling out of time. In her telling, the moon's shadow arrives to the sound of screaming, as if long-buried terrors came rushing up the throats of those watching. As darkness closes over the face of the sun, a shadow also seems to descend over the minds of the crowd, for whom the world becomes dead and distant, home to only fading memories and faint attachments. For as long as the shadow holds sway, it seems as if everyone gathered there had once loved both their lives and the planet, Dillard writes, but can no longer quite recall either. People screamed because of the speed of the moon's shadow as it crossed the valley. It was 195 miles wide, so wide as to seem endless, and travelled at 1,800 miles per hour. And yet the corona itself looked stationary. It was, she records, like a 'still explosion' – like the patient growth of a lichen that is invisible to the naked eye; or images of the exploding Crab Nebula, expanding through seventy million miles of space per day but appearing unchanged in photographs taken decades apart.

Today we live at the centre of a still explosion, caught up in the illusion that everything is as it was. This illusion was given a name in the mid-1990s, when a marine biologist called Daniel Pauly noticed that subtle shifts in perception took place with each new generation of fishermen. Anecdotes and photographs would prove that the size of their catch was shrinking, and yet each generation had a fixed sense of its own yields as the norm, whereas stories of larger catches in earlier generations were dismissed as fishermen's tall tales. What should have been an observable decline was swallowed by rumour. Borrowing a term from landscape design, Pauly called this 'shifting baseline syndrome'. Each generation assumes that their quieter world is what has always been.

There is an eeriness to experiencing the world like this. The cultural critic Mark Fisher defined *eerie* as a matter of emptiness, and where we do or don't expect to find it. 'The sensation of the eerie', Fisher writes, 'occurs either when there is something present where there should be nothing, or there is nothing present where there should be something'. According to the ecologist Jens-Christian Svenning, humans have unknowingly occupied eerie landscapes since long before recorded history. Around the world, he argues, areas we think of as wilderness are haunted by the ghosts of large animals. Whereas now mammalian megafauna (weighing more than forty-four kilos) are limited to niche environments, for forty million years they were richly abundant. Climate change may have played a role in the late Quaternary extinction that arrived around fifty thousand years ago, but the main driver was us – wherever *Homo sapiens* arrived outside Africa, megafauna populations haemorrhaged. They didn't just disappear, however; they left gaps that persist today. Big animals play an important role as dispersal agents for a variety of large-seeded or large-fruited plant life, and it is likely that the same was true of extinct megafauna – until they were all gone.

But evolutionary change takes a long time. Many plant species alive today are only superficially adapted to present planetary conditions; most are primarily equipped to thrive in vanished ecosystems that existed before humans arrived to modify them. The first wave of human-driven extinctions produced profound ecological changes: some plants found their ranges radically contracted because they lost their dispersal agents; other populations, no longer kept in check by grazing herbivores, exploded. The dense forests thought to have once covered much of Europe and eastern North America were very possibly the result of mass extinctions.

The image of the dark, primeval forest – a place of magic and mystery that suggests the possibility of other orders of reality – underwrites many cultural imaginations around the world. Yet perhaps it has remained so eerily compelling because we intuit that this is not how things should be. Maybe we have inherited a sense of plenitude that makes us recoil from the silence, and wonder why there is nothing where there should be something.

CHANGE ISN'T ALWAYS SO SLOW, however. Human-driven evolution is observable in every major taxonomic group (animals, plants, fungi and microorganisms), at rates that far exceed natural processes. Since the development of farming, we have domesticated 474 animal species and 269 plant species. And the selective pressures of the Anthropocene – climate change, ocean acidification, soil and water pollution, invasive species and pathogens, pesticides and urbanization – are urging species to follow new evolutionary paths. Insects and weeds develop resistance to pesticides; farmed fish mature earlier than those in the open ocean; animals have been observed to change their behaviour, body size and colouration to adapt to the arrival of invasive species. Some, in learning to live in urban environments, forget their affiliation with the rest of their kind: *Culex pipiens molestus*, a common house mosquito, has evolved to live only in pools of standing water in the tunnels of the London Underground, and has lost the ability to interbreed with mosquitoes from the surface. In the midst of the gathering silence, we are writing ourselves in the bodies of many of those that survive the onslaught.

One creature in particular is thriving in the world we've made. Dense jellyfish blooms are said to be becoming more common, sometimes packing in up to ten creatures per square metre and stretching for miles, in clotted water so thick you could imagine walking on its surface. In Japan, blooms of Nomura's jellyfish – a giant that can grow to two metres long and weigh up to two hundred kilos, like a floating refrigerator – were once limited to a forty-year cycle but have become a near-annual occurrence. The blooms are so enormous they have been known to capsize fishing boats. It was estimated in 2006 that jellyfish biomass was three times greater than the total biomass of all fish in the world's oceans. Jellyfish are very difficult to study, and much about the causes of blooms and the rhythm they follow is poorly understood. Nonetheless, some marine scientists are worried that the animals may be riding a wave of man-made effects that will carry the planet back to a Precambrian oceanic world dominated by boneless and tentacular things.

Jellyfish (a term that includes over two thousand species of cnidarians and ctenophores) are not actually fish, but they are an immensely ancient life-form. As soft-bodied creatures, they rarely leave behind a fossil, but a quarry in Wisconsin retains rare evidence of a mass bloom of thousands of jellyfish, some up to three feet in diameter, washed up on the shore of what was once a tropical sea more than half a billion years ago. The stranded animals would have been covered in sediment within a matter of hours, preserving them as cup marks in the sandstone – faint shadows, bearing witness to an astonishing evolutionary continuity. Jellyfish have survived, largely unchanged, through virtually every planetary extinction event, including the Great Dying. Some scientists think that jellyfish may have swum in Precambrian oceans up to 640 million years ago, long before the explosion of complex life-forms with mineralized skeletons that, eventually, led to us.

Where other animals must adapt to live alongside us, jellyfish are untroubled, uniquely suited to flourish in environments that increasingly recall the deep past. The Precambrian ocean was much lower in oxygen than it is today, yet all over the world, the seas are suffocating and draining of life. Large marine dead zones edge some of the most heavily populated stretches of coastline, in the Bay of Bengal, the South China Sea and the Gulf of Mexico. Dead zones occur when an increase in the nutrient load in a body of water leads to a depletion in oxygen in the lower parts of the water column, in a process called eutrophication – making the benthic layers nearest the seafloor uninhabitable for many creatures, with implications that ricochet up the food chain.

One of the main causes is a surfeit of nitrogen. For much of the planet's history, nitrogen has been both hugely abundant – constituting 78 per cent of atmospheric gases – and immensely difficult to metabolize, being locked away in forbiddingly tough molecular bonds. Around two and a half billion years ago some forms of microbial life devised a means to use inert atmospheric nitrogen to synthesize nucleic acids and proteins, setting in motion the modern nitrogen cycle. Still, sources of reactive nitrogen remained scarce until 1911, when the German chemists Fritz Haber and Carl Bosch developed a process to produce reactive nitrogen on an industrial scale, combining the inert gas with hydrogen derived from natural gas at immensely high pressures and temperatures. It was the most radical intervention in the circulation of nitrogen in two and half billion years. Between 1960 and 2000, use of nitrogen fertilizers increased by 800 per cent. Combined with oxidized nitrogen belched from car exhausts, human activity has doubled the natural terrestrial rate of nitrogen fixation.

Massive quantities of anthropogenic nitrogen, as well as other kinds of industrial and agricultural waste chemicals such as phosphorus, now find their way untreated into rivers and waterways around the world. According to the chemist James Elser, 'the world is awash with nitrogen in an inadvertent global-scale biogeochemical experiment'. The excess nutrients, which stimulate the growth of phytoplankton at the base of the food chain, initially cause an immense surge of life – everything from tiny copepods to whales arrive to binge on the windfall. But this feast comes at a cost. The algae blooms are too great to be eaten all at once; dead plankton sink to the seafloor, along with the vastly increased amount of waste produced by the grazers gorging on the algal bonanza, prompting a second microbial banquet that consumes most of the oxygen at the bottom. Because the hyper-nutrified water that causes the blooms in the first place is warmer and less salty than seawater, it settles on the surface like a wax seal on a jar of honey. The entire water column stratifies into healthy and unhealthy layers, preventing the mixing that would reoxygenate the lower layers. The result is an area of water largely (hypoxic) or wholly (anoxic) depleted of oxygen, in which very little can live, except bacteria and jellyfish.

The number of dead zones has doubled every decade since 1960. The phenomenon has been observed since the beginning of the last century, but there is no recorded instance of a significant ecosystem recovering once it becomes a dead zone. In 2011, there were more than 530 worldwide.

We may be approaching a tipping point, from ecosystems that favour fish to seas dominated by jellyfish. If we continue to devastate large areas of the oceans, then, like Woolf's heeling boat, at some point the switch will become irresistible, and extraordinarily difficult if not impossible to put right. In light of this, we can see today's blooming jellyfish as a living example of a possible future fossil, bobbing around in overfished waters too acidic for crustaceans to form shells. Where there ought to be nothing, because the conditions for life have declined so steeply, something may persist.

However unlikely it is that a soft-bodied creature will leave a mark in rock, if jellyfish come to rule the oceans again, then we may anticipate a future in which the most likely source of fossil evidence of life in otherwise largely empty seas will be another catastrophic stranding like that left in the Wisconsin quarry.

IT HAD BEEN a fairly bumpy ride, bouncing along on the tops of the waves, and I was glad to have firm ground under my feet again. We docked in the old boathouse, lined with nets and poles and a bright pink bicycle strapped to the wall, and the water under the boathouse roof glowed greenly. Outside, however, the sky was a deep, late-summer blue, and the air was filled with the warm smell of cut pinewood. To the right of the small harbour, three men were constructing a new boathouse to join the small cluster of wooden buildings – all painted in traditional *falu*, a brickish red dye – that make up Stockholm University's Baltic Research Station on the island of Askö.

Askö is a narrow, ten-kilometre-long finger poking through the archipelago south of Stockholm, with the field station huddled in its crook on the western side. Visiting scientists come and go, as their work requires; otherwise, the island is only permanently inhabited by a single farmer and his cattle. I had come with some Swedish colleagues to meet Lena Kautsky, the station's former director, and to learn more about the kind of marine environment in which jellyfish are thriving.

The Baltic Sea is the largest marine dead zone on the planet. Bordered by nine northern European nations, a combined population of eighty million people, and a single outflow to the North Sea that is only twenty kilometres at its widest point, the Baltic Sea is a sump for agricultural-chemical runoff, untreated sewage, discarded plastic, industrial toxins and heavy metals, dumped chemical weapons such as mustard gas, and radionuclides from Chernobyl. A recent report by the Baltic Marine Environment Protection Commission, a partnership of all Baltic nations to protect the region and ameliorate the pollution, estimated that 97 per cent of the sea was affected by eutrophication.

And yet, nothing seemed untoward about the scene before us. Gulls swooped above the shore, and a few ducks bobbed around in the harbour (we later learned that three of them were plastic, marking out the underwater route of a diving trail). The only sound was the occasional knock and tap of the men at work on the new boathouse.

Lena led us through the station to a light-filled upper room perched like a bell tower on the central building. She had recently retired and wore a silver bladderwrack necklace around her throat, an iron pigtail over one shoulder, and a look of inexhaustible patience. The room was lined with windows looking out onto low wooded islands and bald skerries. Behind the boathouse where we had arrived, I could see the station's two research vessels. The larger was named *Electra* (for a species of bryozoan, *Electra pilosa*, not the daughter of Agamemnon and Clytemnestra, Lena said); the smaller was called *Aurelia*, after *Aurelia aurita*, the moon jellyfish.

Askö was one of three stations established in the early 1960s, after outbreaks of what were known in the Swedish press as *mörder algae* (literally, 'killer algae') laid waste to the ecosystems around Sweden's coasts. Increases in nitrogen-fixing cyanobacteria, or blue-green algae, are a seasonal occurrence in the

Baltic. Lena told us there is even evidence, in sediments taken from the seafloor, that phytoplankton blooms were common in the Medieval Warm Period. But they are grossly exacerbated by the effects of nitrogen and phosphorus pollution.

The peculiar features of the Baltic Sea make it especially prone to eutrophication. It is the world's youngest sea, a creation of the last glaciation between twelve thousand and fourteen thousand years ago, but it sits on top of the same ancient bedrock excavated to house Onkalo. Since the glaciers retreated, land elevation and sea-level rise have competed for influence, and the Baltic region has ebbed and flowed between freshwater lake and saltwater sea. Since it assumed its current shape around seven thousand years ago, the Baltic is the second largest body of brackish water in the world after the Black Sea. Combined with the bottleneck effect of the narrow link with the North Sea and the Baltic's sink-like topography (consisting of a series of large basins up to 460 metres deep, surrounded by shallow sills), its brackish composition means that its waters are naturally stratified, with a lighter, less salty layer at the surface preventing the downward mixing of oxygenated water.

Lena explained that the worst period for eutrophication was in the 1970s and '80s. Parts of the Baltic have since recovered. 'Then, I could not swim in the sea around the Stockholm archipelago', she said. 'Today, I can swim there whenever I want.' But restoring an inner coastal area bounded by islands is a very different matter to managing the open sea, she told us: 'The worst affected area of anoxic water, out in the Gotland Deep, is roughly the size of Denmark.'

I asked Lena if the future of the oceans meant learning

to live with dead zones. Good things were happening, she insisted: recent research suggested that better management of sewage and industrial and agricultural waste (at least by some of the Baltic nations) meant that nitrogen levels were declining and phosphorus was plateauing. But the Baltic operates according to its own timescale, outside political imperatives. The single slim outflow means that a full exchange of the Baltic's waters with the North Sea happens very slowly. If we filled the Baltic with red dye, it would take three decades for it all to disappear, Lena said.

She led us downstairs to a wall of fish tanks, which represented the most common Baltic species. They bubbled away contentedly, but seemed oddly empty. Instead of darting fish, the tanks held only eelgrass, bladderwrack and blue mussels. The brackish Baltic waters, for which relatively few species are adapted, make for a naturally constrained ecosystem. Rubbery bladderwrack dominates the shallow reaches where sunlight penetrates, up to depths of twelve metres; blue mussels clutter the shore. These filter feeders can sieve a litre of water an hour; in a year, the population of blue mussels in the Baltic can filter a quantity of water equivalent to the volume of the entire sea. This also makes them sinks for toxins, however, which, as they are picked off by seabirds, concentrates much of the pollution higher up the food chain.

Aside from the eelgrass and the algae, there is very little else. Lena told us that only half a dozen species live in the sediments at the bottom of the Baltic Sea: 'That's why it's so easy to work with!'

The simplicity of the Baltic ecosystem seemed remarkable. This was a natural sparseness, the consequence of a small number of species adapting to fit a particular niche. But it also felt like a warning, a foretaste of what could become of other, more plentifully populated seas if dead zones are allowed to spread. The month before I visited Askö, I travelled to the Sven Lovén Centre for Marine Infrastructure, another marine research station, on Tjärnö, an island on Sweden's fragmented west coast. The North Sea has seen an appalling decline in large marine life – around 97 per cent of fish over four kilos are gone – but compared with the Baltic these waters are still rich in smaller forms of life. Here, too, though, changing conditions represented a threat to the abundance. Lena told us that she has a house on Tjärnö and had often witnessed vast blooms of moon jellyfish as far as she could see.

Tjärnö is a landscape of pine, spruce, and fifty-million-year-old pink-veined granite. As it had on Askö, the sun shone mildly on Tjärnö's wooded islands and sheltered beaches. The Tjärnö station is larger than the Baltic one, and inhabited permanently by a mix of researchers, students and visiting scientists with their families. Children splashed in the shallows as we boarded the centre's research vessel. Everything was warm, blue, and peaceful.

Our skipper was Kerstin Johannesson, a professor of marine ecology with a severe crew cut and a weathered manner. 'This ship is the only place in Gothenburg University where the vice chancellor is not in charge!' she said, grinning, as we made our way out into the Koster Fjord.

In its deepest part, the Koster trench goes down 400 metres, although it has been backfilled over the past 300 million years by more than 250 metres of sediment. Our purpose on this trip was to collect hard and soft bottom samples, to see what kind of small creatures lived on the floor of the North Sea. We were told to expect surprises. 'Don't blame us if we can't identify everything', Kerstin warned; the mud we were to dredge is home to more than six thousand species of marine plants and animals.

Dredging is a curious mix of the very blunt and very fine. The dredge itself looked like a steel milk crate with a net hanging beneath it. When a full sample was winched on board, the contents were dumped onto a wooden pallet and hosed down to wash away most of the shining, velvety mud. Then it was spaded into plastic crates, with a sink on deck for sifting and sorting the tiny specimens by hand.

The first deposit of silky slop revealed a teeming congregation of postage-stamp-size brittle stars, flexing their long, tapering arms. Brittle stars are a kind of starfish. Their arms can regenerate to replace lost limbs, so they play an important part in stem cell research. There were countless separated arms in the tray, curling blindly among our mess of brittle stars like strange, hairy minims. They don't have anything that would approximate to a face or eyes, but I later learned that brittle stars have evolved to see with their entire bodies, or, rather, with their bones. Calcium carbonate crystals in their skeletons work like tiny lenses, focusing light so that the entire animal forms a single compound eye.

With a rush, the second sample crashed onto the deck like a mess of broken crockery: crab limbs, grizzled oysters, shards of storm-coloured mussel shells, bulbous *Alcyonium digitatum* (a form of soft-bodied coral otherwise known as dead man's fingers) and leathery strips of kelp. Digging around in the glossy sediment, another of the marine scientists, Matthias Obst, drew out what looked like a lump of melted plastic. 'This is your relative', he announced. The lump was a tunicate, a species of invertebrate that has a kind of proto-spine during its larval stage. It is supposed that a mutation allowed some tunicate larvae to retain their spines, a development that eventually led to vertebrates like us. The animal wore an outer 'tunic' of scarlet cellulose (hence the name), a soft, pliable shell that resembled cellophane. Using his thumbs, Matthias peeled it away to reveal the shapeless animal beneath, a blob of gristle trembling gently in his palm with the thrum of the boat's engines. 'It's fine', he reassured us as we looked on in alarm, 'it doesn't hurt them. The coat will grow back within three weeks.'

Turning it this way and that, he angled the tunicate toward the light. 'They also have a heart', he explained, 'but it's quite difficult to find.' Their circulation follows a tidal rhythm: flowing seven times in one direction, then reversing. An indignant jet of water spurted out of it (tunicates, I learned, are filter feeders just like mussels), surprising us all, but we couldn't find the heart.

Back at the station on Tjärnö, we hefted the crates of samples onto the deck and carried them up to the laboratory. For all the astonishing finds we had already made in the Koster Fjord's shining silt, it was here, under the illuminated gaze of the microscope, that the fabulous riches of this sea became apparent: brittle stars waving in the limelight; a starfish's nearly transparent limb lined with hundreds of undulating, creamy cilia; the oblate eyes of a crab. I picked up a brownish scrap of leathery seaweed printed with silvery patterns. To the naked eye it looked unremarkable, like mould on fruit or spent bubble wrap, but the microscope showed these grey blushes to be whole cities of lace coral. A petri dish of sea-water revealed a fathomless world of glinting phytoplankton and zooplankton, like a shower of silver coins.

The tang of cold water was in my nose, my fingers soused in a seawater smell. Despite the depletion of the North Sea's fish, this environment was still vital, thriving with uncountable life. Matthias beckoned us over to a microscope where he was disrobing another tunicate. There, among the confusion of its snot-like form, was a tiny red spot, pulsing slowly according to its body's own tidal push and pull.

On Askö, after showing us the simulated Baltic Sea envi-ronments in the tanks, Lena took us to see the real thing. The Baltic coast may be a marine desert, but Lena wanted to show us what life did manage to endure. She led us through the woods to a section of shoreline near the diving trail and the plastic ducks, and as we reached the sea's edge she pulled off her shoes and began fishing in the water with her toes. The lip of the rocks was coated in a thick beard of lime-green algae, a remnant of that year's bloom. 'The algae become greener as more nitrogen is added to the water', she said. 'It's like a paint chart. You can tell how eutrophic the water is by the colour of the algae.' Balancing carefully, Lena withdrew her foot and plucked a clump of bladderwrack from between her toes. Her face lit up, and I saw why she wore her bladderwrack necklace. She had spent her working life studying this kind of seaweed – 'the forest of the Baltic', she called it – and was still full of admiration for this hardy plant that maintained a hold in such a precarious environment.

As we walked back to the station, on a path carpeted by orange pine needles, Lena told me how a restaurant in Stockholm had contacted her about which local jellyfish they could serve to their customers and where to source them. Did you ever eat the jellyfish yourself? I asked. No, came the reply, but Lena had once tried jellyfish pie. I wondered aloud what it tasted like.

'Nothing, it tasted like nothing', she said, 'just salty water.'

THE WAYS IN WHICH we are reshaping the oceans to suit jellyfish and little else are many and varied. Whereas the poor visibility of eutrophic waters works against visual predators like fish, this is no detriment to jellyfish; they also reproduce and grow more quickly, making them better able to exploit the nutrient bounty. Some are even able to absorb up to 40 per cent of the nutrients they need directly through their bodies. Overfishing reduces both the predators and competition that otherwise limit jellyfish numbers. Intensive bottom trawling, raking the seafloor with huge nets, razes the habitats that suit fish, such as coral reefs, and creates instead 'microreefs' – from discarded pieces of net to the tiny broken arms of brittle stars – that are ideal for jellyfish polyps. Trawling also destroys the bottom-dwelling filter feeders, such as clams, that eat the polyps. Added to this is the enormous increase in what is called hard substrate, the firm surfaces that polyps need to attach themselves to. Oil platforms, offshore wind farms, shipwrecks, aquaculture, harbour walls, and marine plastic waste all increase the available store of polyp habitat, often at the expense of other marine creatures. It's thought that the blooms of enormous Nomura's jellyfish are due to the proliferation of concrete embankments along the rapidly developing northern coastline of the East China Sea.

There's also climate change. Jellyfish metabolisms speed up in warmer water, whereas more acidic water softens the shells of many crustaceans; warm water may even trigger reproduction in some jellyfish species. The shipping lanes that knot together the world's ports act as jellyfish highways, as the animals are sucked up in the ballast water of ships and dumped on arrival, stirring up new ecosystems in distant ports that are poorly adapted to cope with these invaders. *Mnemiopsis*, an egg-size jellyfish, belied its diminutive size to spread unstoppably around the world within a few decades; from the United States, it colonized the Black Sea, the Caspian Sea, the North Sea, the Baltic Sea and the Mediterranean.

Jellyfish success is having all kinds of unexpected, eerie consequences. Power plants are usually sited on the coast because they need access to large quantities of water to cool them down. In June 2011, a mass of moon jellyfish blocked the filtration systems of two nuclear power stations, separated by more than 5,500 miles, forcing both to shut down temporarily: at the Torness nuclear power station, fifty kilometres from Edinburgh and within sight of the beach to which I take my students on our field trip into deep time; and at the Shimane nuclear plant in western Japan. This was only three months after the tsunami that destroyed the nuclear plant at Fukushima, when fears of another nuclear disaster were running high. The following month, jellyfish stopped energy production at the Orot Rabin electrical power station in Hadera, Israel, and the same happened again in a nuclear power station in Florida. The Oskarshamn nuclear power plant, on Sweden's Baltic coast, was closed by a mass jellyfish influx in 2013, having also been affected in 2005. In response to this threat to energy production, South Korean scientists have designed robotic 'jellyfish terminators': automated machines that can mash their way through up to nine hundred kilograms of jellyfish per hour, or around six thousand animals.

I wanted to know what a jellyfish bloom was like, so after I returned from Askö I began watching online videos of swarms. If you muted the melodramatic narration and spooky music, they became quite hypnotic, like pink blizzards. In some, the divers were nearly swallowed by white clouds, as if they were in the middle of a foam party. It was easy to lose a sense of scale. Palm-size transparent jellies wheeled slowly like tiny spiral galaxies.

Although they are becoming more frequent, jellyfish blooms are nothing new. In the account of his journey to Spanish America between 1799 and 1804, Alexander von Humboldt described a jellyfish bloom so great that it all but halted his ship in the water and stretched far into the distance like a cobbled road. The slow procession took forty-five minutes to pass, before the ship could move ahead again unimpeded. Such a remarkable sight was one they would come to see often in the Pacific, Humboldt recorded. An 1880 edition of *Nature* reported a bloom of moon jellyfish in Kiel Bay, off the German coast, so thick that an oar jabbed into the swarm would remain upright.

Massed together, individually delicate jellyfish become

irresistible. In 2006, thousands of jellyfish crippled a nuclear warship, the USS *Ronald Reagan*, by flooding the condensers that cooled its nuclear reactor. Singly, however, they seem barely to exist at all. In 1795, Mary Wollstonecraft encountered a swarm of jellyfish in the Oslo Fjord. The sea was calm and clear, and the jellies bobbing just below the surface looked 'like thickened water'. But when one was ladled into the boat to allow a closer inspection, it lost all shape and lustre, slumping into a colourless slop. The Swedish poet Tomas Tranströmer writes of a bloom in the Baltic Sea in the 1970s that resembled 'flowers after a sea-burial'. But he, too, discovered that out of the water, the soft bodies became indistinct, 'as when an indescribable truth / is lifted out of silence'. Jellyfish disappear in our efforts to know them. When we strand them on the shores of our curiosity, they collapse into formlessness.

The difficulty stems from their extreme otherness. Jellyfish have neither a face nor a brain, possessing instead a 'nerve net' spread evenly around their bodies. It's almost impossible to imagine a mode of being so liquid and gelatinous, in which sensation and perception are not concentrated but diffuse. 'What must it be like', writes the poet Jean Sprackland, 'to have no bones, no guts, just that cloudy blue inside you?'

Rather than online videos, hand-drawn illustrations helped me to see jellyfish. In 1864, the evolutionary biologist Ernst Haeckel was in Nice after the sudden death of Anna, his wife of just eighteen months. One day, walking along the Mediterranean shore, he came across a jellyfish in a rock pool. Its curling blond tentacles reminded him of his dead wife's hair, an association that stayed with him even when his grief was blunted by the passage of years. Forty years later, in his great work of zoological illustration, *Art Forms in Nature*, Haeckel named what he thought the most beautiful species of jellyfish *Desmonema annasethe*, to 'immortalize the memory of Anna Sethe', to whom he owed the happiest period of his life.

Art Forms in Nature teems with natural wonders: feathered copepods, diatoms like crowns of bone, and tunicates that gleam like Fabergé eggs. But it was the remarkable beauty and balance of the jellyfish that absorbed me. On every page, all the available space is used with startling efficiency, yet the pages never feel crowded. Some are arranged with a kind of heraldic symmetry: roped, frogged, embossed with gold detail and washed in imperial pink. They made me think of church cushions, Celtic knots, or wax seals: as heavy, elaborate and strictly controlled as Victorian wallpaper designs. But in others, the jellyfish flow elegantly around one another, making it easy to imagine a world of soft collisions, silent predation, and streaming trains of tentacles. Some have a vegetable appearance. The jellyfish that so reminded Haeckel of his dead wife has a fronded bell like a cornflower trailing long pink roots. *Leptomedusa* appears in a series of cold, spectral-blue ink washes, as if the creatures were simply emanations of water. The belled forms, viewed in profile, look like drifting hot-air balloons with the ballast ropes cut.

Some of the most startling illustrations would not look out of place in a B movie. The page of box jellyfish is dominated by a headless angel. In one image, done in ghostly white against a dark background suggestive of oceanic depths, the animal looms nightmarishly into view in the centre of the page, predatory and inevitable. The bell is flattish and rigid looking, with panels like a turtle's shell. Kale-like fronds flow out from the underside, into thick tentacles raised like the fighting arms of a praying mantis.

According to the biologist Lisa-ann Gershwin, jellyfish succeed because they are essentially weeds. They occur in a huge variety of forms and inhabit a range that takes in all parts of the ocean, from the floor to the surface, and every latitude, from chilly polar waters to the blood-warm tropics. Each species has two alternating phases in its development: a benthic (occupying the seafloor) polyp phase and a pelagic (occupying the water column) medusa phase. Each reproduces differently. The medusa, which is the drifting, gloopy, softened-glass form we associate with jellyfish, practise sexual reproduction, bringing together male and female to generate polyps. Rather than maturing into new medusae, the polyps themselves also reproduce, but they do so asexually by a process known as strobilation. The polyp elongates and separates into discs of larval medusae, like a croupier doling out gambling chips. They are, effectively, cloning themselves. Even if the polyp dies, its genetic clones live on to create new medusae.

One species, *Turritopsis dohrnii*, takes this a significant step further. It's tiny, only millimetres across, with hundreds of threadlike tentacles and a transparent bell revealing a vivid crimson spot, its stomach. When the medusa dies, a handful of its cells break away from the dead animal and somehow (no one quite knows how) find one another, clumping together in microscopic colonies that, in time, form a new polyp. All complex life on earth lives in the shadow of death; cell death is the price of sexual reproduction. Almost every species that has existed has gone extinct, perhaps as much as 99 per cent of everything that has ever lived. Uniquely, among all this destruction, *Turritopsis dohrnii* has devised a way to live for ever. Whatever the condition of the future oceans, it will be there, pulsing gently into eternity.

IN JANUARY 1929, not long after the New Year, Virginia Woolf was wrestling with the book that many would come to regard as her most important novel. 'I am haunted by two contradictions', she confided in her diary. 'This has gone on for ever: will last forever; goes down to the bottom of the world – this moment I stand on. Also it is transitory, flying, diaphanous. I shall pass like a cloud on the waves.'

In *The Waves* Woolf came closest to describing her abiding sense that another world, of remembered impressions and sensations, accompanies the present. Her object, she declared, was to create a work saturated by life, by everything – however contradictory – that accompanies 'the moment'. At the very end of the book, she returns to the eclipse she witnessed in 1927. In this version, as in her earlier accounts, 'the earth was a waste of shadow', but here she is most preoccupied with the return of the light – first in frail strips, then a spark, and 'a vapour as if earth were breathing in and out, once, twice, for the first time', light gathering in weight and intensity as the earth absorbed colour 'like a sponge slowly drinking water.'

The horror of the eclipse is temporary; the shadow passes, and life returns. Fragile but inevitable, the spark that leaps from the thin corona ignites the world again. The same will be true for the current extinction crisis. After each past extinction, no matter how close things came to a total end, biodiversity recovered. All animals alive today are descended from the paltry 4 per cent of species that survived the Great Dying, but the lapse of time involved is forbidding. We live in the shadow of an eclipse that will endure perhaps as many as ten million years before sound, shape and colour return in full to the land and the oceans.

In the meantime, gaps are there to be filled. Where nothing ought to be able to live, something will thrive. Jellyfish ruled the oceans for millions of years, and they stand ready to inherit them once again. The sea, wrote T. S. Eliot, has 'many gods and many voices'. But perhaps not for long. In the future, if all other light and colour is extinguished from the seas by our carelessness, the ocean's one lonely god will be frilled and eyeless, drifting placidly and implacably through its vast, empty dominion.

EIGHT

THE LITTLE GOD

This is how the Roman poet Ovid imagined the world's making. Before there was anything, there was everything: a roiling mass of matter without shape or edge. All was dark, and in the darkness the elements clashed ceaselessly – hot with cold, pliant with strict, wet with dry.

Order arrived with a god, who parted the earth and the heavens, wrapped the waters around the land, and set the clear air apart from both. The highest sphere consisted of the weightless, fiery ether, which formed the vault of heaven; beneath it was the denser air; finally, heaviest of all, so heavy it sank beneath its own mass, was the earth itself. The god shook out the land like a wrinkled cloth and put the seas around it to keep it in place, fencing the winds to keep them from tearing the world apart again. So appointed, it became a home for many different creatures, each adapted to its zone: fish in the waters, wild beasts on the land and birds in the air.

Like the god in Ovid's *Metamorphoses*, microbes engineered the conditions for life to prosper. The oxygen-rich mix of atmospheric gases that sustains complex life is the product of photosynthesizing cyanobacteria, which began to form colonies around 2.7 billion years ago, when there was only 1 per cent as much free oxygen in the atmosphere as there is today but one hundred times as much carbon dioxide. Over the course of several hundred million years, cyanobacteria raised the oxygen content of the atmosphere to within 10 per cent of today's level, supercharging the evolution of complex lifeforms. Microbes are the authors of every key process of life: fermentation, photosynthesis, cellular respiration and the nitrogen cycle. Their potential is prodigious. According to the sociologist Myra Hird, just one cyanobacterium left to multiply on a sterile planet could oxygenate the atmosphere within forty days. Microbes invented metallurgy and communal living; they are the progenitors of the planet's energy cycle, having devised means to derive energy from solar, chemical, organic and inorganic sources. Perhaps as many as fifty billion species of multicellular life have swum, walked, flown, burrowed or crawled over, across and through the earth, and every single one has owed its existence to microbial ingenuity. Without microbes to assist decomposition, the world would be no more than a heap of the dead. We could say that the biosphere as it is today is one immense microbial footprint. To borrow Ovid's fabulous phrase at the start of *Metamorphoses*, microbial life has spun 'an unbroken thread of verse, from the earliest beginnings of the world.'

My search for future fossils brought me to one of the biggest cities, the largest dead zone, and the greatest living structure on Earth. But as I thought about these most massive and visible footprints we are leaving behind, about the continent-spanning road network and the country-size gyres of plastic swilling in the oceans, I began to wonder, too, about the very smallest. Many of the key life processes authored by microbes now also carry our imprint. Our intervention in the nitrogen cycle is the most significant impact since early microbial life began fixing inert atmospheric nitrogen two and a half billion years ago. We have accelerated the natural rate by 100 per cent in just one hundred years, leaving telltale chemical traces in sediments and plant fossils, and in marine environments devastated by nitrogen and phosphorus pollution. We have fundamentally altered the carbon cycle, shifting billions of tonnes from the slow, multi-millennial cycle that moves carbon between the atmosphere and the ground beneath our feet to the fast cycling of carbon through the bodies of dead plants and animals. Before the last of our carbon traces have been drawn down into rocks and oceans, the heating effect of the extra carbon will have remade the world, altering marine chemistry, flooding coastlines, stripping glaciers to bare bones, emboldening deserts, warping the circulation of ocean currents, supercharging extreme weather events along coastlines and fires on land, and rearranging the distribution of animal, plant and microbial species across the globe.

Cyanobacteria even changed the colour of the world, adding oxidized reds and umbers to the palette of greys and greens, and you need only consider satellite images of the ashy sprawl of our cities or the parched yellow of desertifying landscapes to see we are on our way to achieving the same.

Microbes are the source of many of the world's minerals. The dust cloud that formed Earth consisted of only 12 of them; several billion years of geochemical turbulence brought

the number of minerals up to 1,500, but the activities of microbial life nearly tripled it in only a few hundred million years.
Yet it has taken just three hundred years of industrial activity
to synthesize another 208 new minerals – mostly short-lived
novel minerals associated with mining – as well as hundreds of
thousands of mineral-like synthetic substances that are much
more apt to survive for a long time; some, geologists speculate, may never have existed before anywhere in the universe.
We have separated metals into elements that have either never
occurred naturally in their pure form, such as sodium, calcium and potassium, or occurred only in trace amounts, like
aluminium, titanium and zinc, producing hundreds of thousands or even millions of tonnes of annually. Our towns and
cities and road networks are composed of immense quantities of mineral-like bricks, concrete and glass, added to which
are masses of equally hard-wearing plastics, silicon chips in
smartphones and laptops, and tungsten carbide in machine
tools, lightbulb filaments and ballpoint pens. We have released
substances like iron and gold from the achingly slow processes
of subduction and weathering and scattered them across the
surface, locked up in the frames of our buildings or buried in
homes, bank vaults and graves across the world.

We have even spread our mineral traces throughout space.
Since Sputnik in 1957 there have been nearly 5,000 rocket and
satellite launches, which, through redundancy and collisions,
have produced an orbiting cloud of space junk. Of the 6,000
or so tonnes of this material – much of it aluminium and plastics such as Kevlar – only around 5 per cent are functioning
satellites: of the rest, NASA is able to track 22,000 objects that
are greater than ten centimetres. But there are also thought

to be between 500,000 and 750,000 fragments of between one and ten centimetres, as well as more than 135 million microscopic particles, mostly chips of paint, travelling in orbit with the force of supercharged bullets at 28,000 kilometres per hour. Regular collisions mean that the number is increasing exponentially. The higher the orbit, the longer the material will last: debris below 1,000 kilometres can persist for hundreds of years; objects at around 1,500 kilometres may stay in orbit for thousands. But those that occupy a geosynchronous orbit, where the Earth's gravity possesses only one-fiftieth of the pull exerted on the surface, can remain there for millions of years. The Laser Geodynamic Satellite, launched in 1974 to track the movement of continents, is expected to remain at an orbit of 5,900 kilometres for the next 8.4 million years.

The six Apollo moon landings left hundreds of artificial objects on the lunar surface – Apollo 11 alone left behind 106 items, including two golf balls and a gold olive branch. More than seventy lunar vehicles have been abandoned on the moon, and because the moon has no weather to erode them, both they and the miles of tracks and 'roads' they made on the surface will persist perhaps indefinitely – certainly, far longer than any road on the surface of Earth. It's thought that the footprints of Neil Armstrong and his fellow Apollo astronauts will be preserved for tens, possibly even hundreds, of millions of years.

Beyond our own planet's orbit, we have landed remotely controlled probes on Mars, Venus, Saturn's moon Titan, the comets 9P/Tempel 1 and 67P/Churyumov-Gerasimenko, and the asteroids 25143 Itokawa and 433 Eros. It's possible that Jupiter's atmosphere is polluted with trace amounts of plutonium and iridium from the Galileo probe. To date, four objects of human origin have left the solar system: the Pioneer (10 and 11) and Voyager (1 and 2) spacecraft. All four bear messages devised in part by Frank Drake, an astrobiologist who also contributed to the WIPP warning system. The designers of the Pioneer plaque, which includes human figures and a diagram of the solar system, called their efforts interstellar cave painting and speculated that they could be preserved in space for billions of years. Both Voyager probes, which could last just as long, carry an eternal library of sounds and images on twelve-inch gold-plated copper disks, including photographs of the Great Barrier Reef, an Antarctic expedition and a modern highway in New York State, and a recording of human footsteps.

At the end of his creation story, Ovid's god made the last of the animals to stand upright, 'bidding him to look up to heaven, and lift his head to the stars. So the earth, which had been rough and formless, was moulded into the shape of man, a creature till then unknown.'

THE FIRST PERSON to see microbial life had to look up to do so. In 1674, using a homemade lens so powerful that it was capable of twenty times greater magnification than Galileo's telescopes, a draper called Antoni van Leeuwenhoek peered upwards through his microscope (like a telescope, because his lenses relied on natural light) and caught the first glimpse of the microbial world in a drop of lake water. The tiny beings were 'diverse colours', he reported, 'some being whitish, others pellucid; others had green and very shining scales'.

More astonishing than their diversity, however, was their size: he estimated that 'ten thousand of these little Creatures do not equal an ordinary grain of Sand in bigness'.

A few years later Leeuwenhoek scraped plaque from his teeth with a quill and was amazed to find that countless tiny organisms even lived inside the human mouth – many more, he speculated, 'than there are men in the whole kingdom'. This would have been an outrageous claim in the late seventeenth century, but it undershot by several orders of magnitude. The average human mouth actually contains one thousand times more bacteria than there are people alive today.

In January 1969 W. H. Auden read an article by Mary Marples in *Scientific American* describing how, to the microbial communities that colonize the surface of the human skin, our bodies must seem like a planet of varying ecosystems – from the 'sparsely inhabited deserts of the forearm' to the 'tropical forest' of the armpits, and the 'cool, dark woods of the scalp'. Auden was so entranced by Marples's vision of the 'cutaneous world' that he wrote a poem, 'A New Year Greeting', addressing his own microbial populations like a god speaking from the heights of Mount Olympus – an alternately benign and capricious god, supplying warmth and shelter but also inflicting twice-daily cataclysms each time he dressed or washed.

Auden's fable gestures toward our tendency to put ourselves at the centre of the grand stories we spin. Human activity has wrought changes to Earth systems that, in scale and significance, are unparalleled since the advent of photosynthesis, and it's easy to let this go to our heads. 'We *are* as gods and might as well get used to it', wrote Stewart Brand on the cover of the first *Whole Earth Catalog*, a revolutionary innovation in grassroots activism published in 1968. Brand's statement – adapted from a remark by the anthropologist Edmund Leach – encapsulated the ambition and the hubris of the second half of the twentieth century, when it seemed that the explosion in technological development had eliminated all barriers to human domination of the planet. Lately, it's been adopted by the proponents of what is sometimes called 'the good Anthropocene': the notion that our genius for technology will ensure that we can continue to pursue energy-hungry lifestyles. As benevolent stewards of the planet, so the thinking goes, we can be trusted to wield the awesome power we have assumed wisely and fairly.

This is as absurd, even cruel, as it sounds. As the ethical philosopher Clive Hamilton puts it, the good Anthropocene's message to those already caught in the grip of drought or sea level rise is stark: 'You are suffering for the greater good'. The Bikini islanders were told something similar by the US military – that evacuating their island home would be for the good of all mankind – and this particular instance of playing god left a corner of the world unfit for life for more than twenty millennia. The fact that we have altered the Earth's geochemical cycles doesn't mean we're in control, but then the Anthropocene isn't about mastery at all – rather, it highlights our intimate relationship with the planet's future, as strange as that which binds us to our microbial communities.

Leeuwenhoek must have felt that he had discovered a world so small as to be entirely separate from our own, existing silently and secretly alongside us, but even Auden's god's-eye exhortation of his own intimate microscopic inhabitants doesn't capture the depth of our entanglement in the microbial world. We can't know ourselves apart from our microbes. Our bodies contain more microbial cells than human cells, performing essential functions linked to metabolism and immunity, growth, even shaping our moods. We have co-evolved with our microbial communities for millennia, shaping and being shaped by one another in ways that have left indelible marks on our bodies and the environments we move through.

Recent research also suggests that we can track disturbances in microbial populations in tandem with each of the great advances in human civilization: the rise of agriculture, the industrial revolution, and the post-World War II acceleration in consumption and technological innovation. According to the microbiologist Michael Gillings, human societies have intervened in the distribution, the population diversity, and even the evolution of microbes. Thousands of years of domesticating plants and animals have also produced domesticated microbial symbionts, as well as human illnesses that originated as animal pathogens. The industrialization of agriculture has created boom environments for methane-generating and nitrogen-fixing organisms. In addition to accelerating the distribution of animals like chickens and the inadvertent spread of jellyfish polyps, the advent of maritime transport allowed diseases such as syphilis, smallpox, bubonic plague and influenza to migrate between continents; microbes travelled across oceans in the bodies of sailors and the ballast water of their ships, to found colonies in new worlds. Biologists refer to contemporary shipping networks as a 'functional Pangaea' for diseases in wildlife, effectively reconstituting a single planetary continent, a geological reality that hasn't existed for 175 million years. Air travel has accelerated the speed of this transfer exponentially, and as climate change advances, the range of microbiotic life will be altered even further, as soil temperatures change or the contracts between microbes and mutualistic organisms like coral are broken.

We haven't only usurped the role of microbes in shaping the atmosphere and generating new mineral compounds; we're also changing the microbial world itself. We are leaving a long-lasting impression on the world we can't see just as much as on the one we can. DNA decays too rapidly to leave a direct fossil, but that doesn't mean there won't be any footprints of the microbial Anthropocene. This is a story of extinctions wrapped within extinctions, and interventions into evolutionary time itself; of efforts to resurrect ice age landscapes and creatures and to write an ancient lament in the basic material of life.

I decided I needed to speak to the scientist who saw our smallest footprints through a microscope.

THE DEPARTMENT OF BIOLOGICAL SCIENCES at Macquarie University in northern Sydney is located in an unprepossessing rectangle of mushroom-coloured brick. Wintry light flooded through the tall windows, and my footsteps echoed softly on the laminated floor. The walls were lined with scientific posters and papers pinned onto notice boards or the closed faces of brown office doors. Around the corner from Michael Gillings's office, my eye was caught by a blast of colour. A random assortment of primary-coloured plastic objects – children's toys, lemon squeezers, shapes I couldn't identify – had been suspended in a cluster in one of the window alcoves. The objects seemed twisted around one another in a maze of pleats and creases that looked like a scaled-up model of a folded chromosome of DNA.

I was reading an article pinned outside Michael's door, about the discovery of up to one hundred million antibiotic-resistant genes per gram of mud in Chinese estuaries like the one I had visited in Pudong ('that's a million resistance genes in a fragment of mud the size of a match head'), when I heard footsteps coming down the corridor and turned to see Michael walking towards me with his hand outstretched. He wore a kaffiyeh and a froth of grey stubble, and my first thought was that he looked more like a surfer than a scientist.

As we sat down in his office, comfortably stuffed with piles of papers and a standing desk surrounded by photographs of sculptures and curious designs, he leaned forward and invited me to fire away.

I asked about the parallels between human effects on the macro and micro worlds. Were they really so clear?

'Yes, absolutely', he said. 'One of the reasons I began thinking about this was through wondering whether we can find out about the microbiology of the deep past by examining what we see now – is there a microbial signature of dinosaur extinction, for instance?'

He leapt up to his desk and brought up a photograph on the computer. The blank screen lit up with vivid colours in undulating bands of blood red and plum black. It was a banded iron formation from Western Australia, he said, laid down around two billion years ago by the action of cyanobacteria. 'In fifty million years,' he went on, 'if cockroaches evolved to become palaeontologists and dug down looking for evidence of what was going on in the early twenty-first, they might find distinct signatures similar to this in relation to the nitrogen cycle. Not that they'd be able to see it by the naked eye, but they'd be able to read the chemical legacies'. So, no colours? I asked.

'No', he replied. 'They'd need to undertake a chemical analysis, but they could also infer microbial changes through associations with other traces like the distribution of phosphorus or plastic pollution. It'd be like looking through a glass darkly.'

He led me out of his office to the floor below. 'So I began to imagine myself as a palaeontologist of the future', he said, his voice echoing in the stairwell. 'What would I find?'

We arrived in front of a large poster behind a glass cabinet. Indistinct burned-looking objects were displayed on small wooden hexagonal mounts. Michael explained that he had set himself and his students the task of making an exhibit of an 'Anthropocene layer' discovered fifty million years from now. Purporting to be the work of 'Hive Consortium 14255', a commune of evolved bees, the poster was titled EVIDENCE FOR A TECHNOLOGICAL SPECIES AT THE ANTHROPOCENE BOUNDARY LAYER. The bee scientists claimed to have discovered definitive sedimentary evidence of a long-vanished species that showed significant technological advancement, including 'unusual minerals and a high concentration of metals and vitreous deposits'. The finds, some of which were exhibited below, included metal tools and 'primitive ceramic representations for extinct animals', and should serve as 'a warning for all hive-kind'.

The source of the Anthropocene layer was a real site in

a nearby park that Michael had taken his students to visit. The photograph on the poster showed tufts of white plastic sprouting from the sedimentary litter. 'We can expect very little change to intra-terrestrial microbiology, the organisms that live in the bedrock', he explained. 'What traces there are will be at or near the surface. Our landfill would be their goldmines.'

Modern landfill sites are designed to keep everything in and let nothing out. Sealed with layers of clay and plastic to prevent toxic materials from leaching into groundwater, they represent controlled chemical environments in which moisture, temperature, oxygen content, and pH levels will remain steady and predictable, slowly curing their contents. The sealed-in leachates, rich in calcium carbonate from buried concrete, may eventually produce a kind of cement to bind the clutter of household rejects, fly-ash particles and soiled packaging into a solid mass. With only a finite supply of oxygen, the deposits will be largely protected against microbial decay. But, as I found out later, a sample of salt crystals taken from a drill site at WIPP in Carlsbad holds out the startling possibility that landfill sites will also contain microbial 'fossils'. The WIPP samples were found to contain living salt-eating bacteria that had survived being trapped in the crystals, 250 million years ago.

The most likely scenario, though, is that the evidence will have to be interpreted. 'The thing about the microbial story', Michael continued back in his office, 'is that we can see it unfolding in real time, just as we can see the extinction of megafauna unfolding. Human domination of the biosphere extends to the microbial world – both are shrinking.' He pulled up a graph that showed how the total biomass of wildlife is dwarfed by that of humans and domesticated vertebrates.

'This is the world's scariest graph,' he said, 'but what few people appreciate is that it also describes a microbial catastrophe.'

Each species has its own unique internal ecosystem, but this is partially determined by the kinds of environments each animal moves through and the other creatures they interact with. 'Imagine you're a lion in a zoo', he said. 'Ordinarily, you would be exposed to new microbes every time you ate a dead animal on the savannah or from the other animals that shit and piss in the watering hole. That's all gone in captivity – now you're eating sterilized meat and only exposed to humans.' A decline at one level leads to a decline at another; when biodiversity drops in the macro-world, it also drops in the microbial. In the final instance, the signs of former microbial diversity will be there in the thinning of the future global pangenome. If the Great Barrier Reef does collapse, then the gap in the fossil record will also mark a microbial calamity. Matryoshka-like, each mammalian or insect extinction contains countless other extinctions or near extinctions as an animal's microbiota lose their habitat.

'All that is left', Michael said, 'is the microbiomass associated with humans and what they like to eat: pigs, cows, sheep, and chickens.'

There was a knock at the door, and Michael's colleague Sasha Tetu came in. Michael had a meeting to attend, so we agreed to finish our conversation later, and in the meantime, Sasha and I went to get a cup of tea out in the sun-washed courtyard.

She told me her research focused on marine photosynthesis and toxicogenomics – essentially, she looked at what conditions would prove fatal for marine microbes. It all started with a very simple experiment, she said. 'I added plastic leachate to a beautiful emerald-green culture of cyanobacteria, and it just bleached it – all the colour drained away to this sickly, pale sludge. It was clear the colony was dead'. She began experimenting with different plastics – PVC, it turned out, was especially toxic. There's so much plastic in the oceans already, I said. What are the consequences for all the microbes out there? 'It may affect the ability of key bacteria to make oxygen – I'm not talking about sterilizing the oceans', she qualified quickly. 'But there could perhaps be a shift in the structure of bacterial communities. I expect the bacteria will adapt, maybe with stronger, weedier variants. That's why I got into microbiology in the first place, because I never cease to be amazed by the adaptability of microorganisms.'

THE PRINCIPLE OF CHANGE is the most essential feature of the microbial world. Single-celled life-forms have been around for perhaps four billion years, and each numberless generation has borne a slight difference from the last, a tiny flaw in the copy. It has even been suggested that there are as many different species as there have been individual bacteria – a notion that is remarkable enough before you consider that there are somewhere in the region of 5,000,000,000,000,000,000,000, 000,000,000 bacteria on Earth at any one time. Another quintillion – 1,000,000,000,000,000,000 – are thought to live on dust particles in the atmosphere, and between 50 and 90 per cent of ocean biomass is believed to consist of microbial cells. Ocean-dwelling microbes are even evolving to consume plastic debris, and in 2016, Japanese scientists discovered a bacterium in a plastic bottle recycling facility that had learned how to break down PET.

The source of this incredible diversity is in their capacity to share their DNA. Through a process known as lateral gene transfer (LGT), any bacterial cell can swap its genetic information with any other cell. Microbes that brush against each other may exchange genes. Some engage in trade; others smash and grab. Or they elect to cohabit, and like long-married couples finally merge to become nearly indistinguishable from each other. Not every exchange results in meaningful adaptations, and unhelpful amendments are often washed out in subsequent generations. Nonetheless, LGT allowed microbes to develop key life processes such as photosynthesis, catabolism, and symbiosis. As in the *Metamorphoses*, microbial life is a continual riot of transformations.

Increasingly, however, our actions are censoring this free expression, curbing the carnival and forcing microbial evolution to follow our lead. Human populations have had an impact on what genes are selected in their own microbial populations via dietary choices, since at least thirty-five thousand years ago, when we began to use fire to cook our food. This 'soft' selection has shifted to a process of 'hard' selection, however, via the use of antimicrobial compounds – at first tentatively, since we began to exploit the prophylactic effects of mercury and arsenic in the late 1800s, but with aggression since the use of antibiotics became widespread after World War II. Overuse of antibiotics is blamed for the steep decline, in developed world populations, of *Helicobacter pylori*, a gut-dwelling microbe that has coevolved with humans for at least one hundred thousand years to assist in regulating the production of stomach acid. And the effects of such antibiotic largesse can be seen far beyond the human gut.

Penicillin was famously discovered by Alexander Fleming in 1928, when he returned from a holiday to find that a petri dish left open on his laboratory bench had been contaminated. The dish contained a large, bright colony of mould and numerous smaller bacterial colonies that reduced significantly in number and size, almost to the point of transparency, nearest to the fungal colony. The mould peeled away the bacteria's cell walls like the shells of boiled eggs. Photos of the petri dish look rather like the moon and stars viewed through a weak telescope. Fleming himself referred to the affected bacteria as 'ghosts'. The drug was finally synthesized in 1943 from mould procured from a rotten cantaloupe, and every variant of penicillin in use today derives from that single sample. Previously untreatable infections could now be resolved by a simple course of pills. Unheard-of procedures, such as joint and organ transplants, became commonplace. Antibiotics have surely contributed to the shaping of future fossils simply by their effect on mortality rates, but they have also had a fundamental effect on which bacteria thrive very far indeed from the bodies of treated humans or animals. They are commonly used to promote growth in livestock and, as between 30 and 90 per cent of the antibiotics ingested by both animals and humans are excreted unchanged into soils and waterways, the microbial world at large is flooding with antimicrobial compounds. In one of his studies of the microbial Anthropocene,

Michael Gillings predicts that this might even affect 'the fundamental tempo of bacterial evolution'.

Resistance to lab-made antibiotics originated at some point in the 1920s, with a single DNA element – the class 1 integron – that has since propagated through LGT in numbers that are difficult to grasp. Millions of copies of the original integron now exist in every gram of human or livestock faeces; up to 100,000,000,000,000,000,000,000 copies are shed into the environment every day, flushed through wastewater-treatment plants (which act as 'hot spots' for LGT, like a kind of supercharged bacterial orgy) into rivers and oceans or fed back into the soil. Class 1 integrons have been found in the Amazon rainforest and in both the Arctic and Antarctic. We have saturated the biosphere with compounds that are accelerating the basal rate of microbial evolution, raising the ratio of genes that confer resistance within the pangenome, the sum of all the genes in every genome of every living thing on the planet. While many of the selective events that produce resistance will be transient, Michael speculates that some changes to the composition of microbial communities may be permanent.

Each gram of soil contains billions of microbial cells. In the right conditions, extracellular DNA can survive in soil or clay for thousands of years, haunting the humus until the moment when it brushes up against a receptive bacterium and, like a spark to a fuse, sets a new line of bacterial evolution in motion.

ON OUR WAY BACK to the department, Sasha asked if Michael had told me about the leaf sculptures. Every autumn, she said, he raked the leaves in the courtyard in the middle of the biological sciences building into a piece of land art. She took me to see the latest installation. 'It's got a bit ragged at the edges', she said, but the outline was still clear: tapering spirals of crisp brown leaves curling out from the base of a tree like the arms of an octopus.

I realized that I'd seen it already, in some of the photos pinned around Michael's office. When he had finished his meeting, I asked him about the sculptures. 'I've been doing it for ten years', he said, 'but I kept it a secret for the first six.' He'd get up early, before the campus was awake, and rake up his design – the idea, he said, was to nudge his students to think about structure and complexity in the biosphere. 'I wanted them to ask themselves, How did this happen? And what changes take place as it decays and merges again with the earth?' The title for this year's sculpture was *The Kraken Wakes*.

There was one more aspect of our footprint in the microbial world I wanted to discuss. The essential rule of biological life has been fixed for almost four billion years – biological information flows from DNA to RNA to protein to phenotype in a closed cascade that scientists call the central dogma. It's immutable, and has been so since the last universal common ancestor of all life. But the development of technologies which lock away digital information in molecules of DNA has broken and extended the central dogma for the first time in 3.7 billion years.

'For the first time', Michael said, 'we have an organism capable of directing its own evolution and the evolution of every living thing on the planet.'

He leapt to his computer again and brought up a series of slides. 'The first chemically synthesized genome was produced in 2007 at the J. Craig Venter Institute, in Maryland. They took the DNA from one bacterium, *Mycoplasma mycoides*, and implanted it into cells of another bacterium called *Mycoplasma capricolum*. The result looked and behaved exactly like *mycoides*'. A few years later Venter went on to synthesize living cells of *M. mycoides* based on information stored on a computer.

Michael picked up a thumb drive and waved it around. 'It's a serious evolutionary transition. Previously, every living thing needed an ancestor; now, they don't. You could upload all the genetic information about a creature here and transport it electronically around the world in seconds'. In theory, he explained, extending the central dogma could allow the reconstruction of extinct pathogens or species, even the synthesis of novel organisms, any one of which would, if it thrived, represent a line in the sand of evolutionary time.

Some have even seen in this new technology the opportunity to resurrect extinct species and remake lost ecosystems. In 1989, Sergey Zimov, the Russian geophysicist who first calculated the vast amount of carbon stored in the Siberian permafrost, established Pleistocene Park, a 160-square-kilometre reserve in north-east Siberia, to test his hypothesis that the restoration of a grass-tundra ecosystem would slow or even prevent permafrost from thawing. During the Pleistocene, Earth pivoted between multiple glaciations, grinding the landscapes of the northern hemisphere beneath immense molars of ice. But the grasslands of Siberia were unscathed, Zimov states. Enormous herds of bison, musk ox, wild horses, reindeer, moose and mammoths roamed across a million square miles of tundra, their busy mouths cropping the grass and keeping the plains free of trees. Over tens of thousands of years, dust blowing across these plains accumulated in deep deposits along with dead plant and animal matter and microbes, which froze into a form of carbon-rich permafrost called yedoma. Zimov calculated that the frozen yedoma deposits alone contain around five hundred gigatonnes of carbon; another four hundred gigatonnes are sequestered in non-yedoma permafrost, and between fifty and seventy gigatonnes in Siberian peat. In all, that's more than four times the amount of carbon sent into the atmosphere through chimneys and exhaust pipes since 1750.

Zimov realized that the yedoma deposits weren't entirely secure. As patches thawed in the summer, forming small lakes, anaerobic microbes in the softening permafrost converted the carbon into methane. The permafrost was exhaling up to forty grams of greenhouse gases per cubic metre every day, fast enough, he estimated, to have expended most of the carbon reserve by the end of the century.

Grass, he felt, was the solution. Bright grasslands absorb less heat from the sun than the dark forests that spread across the Arctic plains, and in winter the large grazing herd animals would break up the snow that would otherwise insulate the permafrost from the harsh Arctic air. Chilled through the winter by trampling hooves, it would be less likely to thaw during the summer. So Zimov has filled Pleistocene Park with as many large herbivores as he can lay his hands on – wild horses and bison roam this small corner of the tundra just as they did twenty thousand and more years ago. But he realized that to really reinstate a Pleistocene habitat, and keep the grasslands free of trees, he needed larger beasts, like those that shaped the Pleistocene plains the first time around. He needed mammoths.

In 2014, a Harvard geneticist named George Church, inspired by Zimov's ideas, began a project to resurrect the mammoth. He calculated that of the 1.4 million mutations that separate the genomes of mammoths and Asian elephants, 2,020 affect protein-expressing genes. Using CRISPR-Cas9, a technique that allows scientists to guide modified proteins into specific points in a DNA sequence, Church has so far cut and pasted into the Asian elephant genome forty-five mammoth genes, harvested from remains frozen in the Siberian permafrost, including information needed for hair, smaller ears (to minimize heat loss), and subcutaneous fat. A cell of Asian elephant DNA speckled with bits of synthetic mammoth is not a mammoth, but Church's gamble is that the effect on the Siberian ecosystem would be the same.

But Michael wasn't convinced that manufacturing a new genome would be enough. Like elephants, mammoths would have had a complex social structure. A new herd would exist in a social vacuum, a world that has entirely forgotten what it is like to be a mammoth. 'Animals have to learn to be animals', he said. 'You can't synthesize behaviour.'

De-extinction would be the greatest god-trick of all, but it won't be achieved just by confecting disappeared species in a laboratory. Animals aren't simply the sum of their parts: each physical characteristic is the result of a host of evolutionary adaptations passed down through thousands of generations. And we would need to change too. Bringing back species lost to the expansion of human dominion would be miraculous, perhaps even an atonement for past transgressions, but it would also require us to learn to live alongside what we once forced from existence.

As Michael walked me out of the building, we paused at a window so we could look down at the leaf sculpture in the courtyard. Despite the frayed edges, the curling shapes were still clear. I asked if he had an idea for next year.

'There was once an artist in Sydney', he said, 'who would chalk *eternity* on the sides of public buildings and on paving stones. I think next year I might arrange the word *entropy* in big copperplate letters.'

MICROBES HAVE EXISTED for billions of years by virtue of their capacity for invention. They are the world's great improvisers, and their ability to evolve beyond their apparent limits seems like a good model for learning to live in the world we've made for ourselves. Microbes show that what we could be is not limited by what we are, and that embracing collaboration and adaptation, shrugging off old shapes to accept new ones, can bring an increase of life.

Their adaptability means that microbes have a tremendous capacity to endure. Bacteria have been found thriving in the deepest parts of the ocean and far beneath the surface of the earth. The record is currently five subsurface kilometres, although some scientists speculate that, given that the temperature limit for life is 122 degrees Celsius, microscopic life could be found twice as far down. It's thought that 70 per cent of all the planet's bacteria and archaea exists in the subterranean darkness. Some even live high above our heads. In

1978, Russian scientists used meteorological rockets to bring back samples of microbial life, including a species of penicillin, from over sixty kilometres into the mesosphere.

Some microbes don't just live hard; they live long. In 2010 Charles Cockell, an astrobiologist at the University of Edinburgh, discovered that a petri dish that had been left undisturbed for ten years contained a sample of desiccated but viable bacterial spores. To find out just how long microbial spores can persist in a dormant state and still revive, Cockell and a team of collaborators designed the 500-Year Microbiology Experiment. They sealed dried samples of two bacteria, *Chroococcidiopsis* and *Bacillus subtilis*, in eight hundred glass vials, locked them in two sturdy oak chests, and set a centuries-long schedule for testing their resilience. Every 2 years for the first 24 years of the experiment, and then every 25 years for the next 475 years, a scientist will remove a vial from each oak box and examine its contents to see if the spores will flourish. The final vial will be tested in 2514. Weirdly echoing the efforts to guarantee a message to the deep future that surround WIPP, the team has had to frame the instructions to their future colleagues – most of whom have yet to be born – with extreme care, in order to safeguard the experiment. At present, the instructions are stored along with the samples in written form and on a USB stick. But like Sebeok's atomic priesthood, each generation of scientists will be charged with making a new copy using the most up-to-date technology and accounting for linguistic changes.

Microbes can bide their time, waiting for the opportunity to flourish, in the least promising of circumstances – even in between the pages of a book. In the same year Charles

Cockell discovered the petri dish with its ten-year-old sample in Edinburgh, an artist called Sarah Craske found a 275-year-old copy of *Metamorphoses* in a junk shop in Kent. She bought it for three pounds, and only later learned that it was one of just three surviving copies of this particular edition. However, it was neither the book's rarity nor even, exactly, the lines of poetry themselves that had interested her, but what had been printed invisibly on the pages: nearly three centuries of microbiological history.

Sarah saw that her copy of *Metamorphoses* was also a library of biological information, and she began working with a microbiologist named Simon Park to discover the book's secret history. They cut out pages and pressed them into bioassay dishes filled with a molten blood agar for twenty seconds; the pages were then removed, and the dishes incubated for a week. This bacterial printmaking revealed the incredible diversity of microbial life that had been wiped or coughed onto the book's pages by generations of readers. Hundreds of colonies bloomed in the plates, including (predictably) many kinds of microbes that live on human skin – one of which, *M. luteus*, was instrumental in Fleming's discovery of penicillin. But there were also more remarkable survivors. Living in the margins of the stories of Actaeon and Narcissus were *Bacillus altitudinis*, which was first discovered in 2006 in air samples recovered from forty-one kilometres above the earth, and *B. subtilis*, used in the 500-Year Microbiology Experiment, which can withstand desiccation and temperatures up to 53 degrees Celsius. A colony of *B. subtilis* survived six years in space on a NASA satellite.

Microbes with this capacity to endure are called extremophiles. They're just about the hardiest life-forms on the planet.

Some species can survive frozen in ice, or in temperatures above 120 degrees Celsius, clinging to the nutrient-rich slopes of hydrothermal vents. Other extremophiles can tolerate immense doses of radiation, or impacts at crushing velocity; even, as demonstrated by *B. subtilis*, in the vacuum of space. Because of extremophiles' incredible resilience, scientists have begun to explore using bacterial DNA as preservation machines for digital information.

Since the rise of data streaming services, artificial intelligence, cloud storage, social media platforms, and the ubiquity of smartphones and watches, it now takes only a couple of days to produce five billion gigabytes of digital content, equivalent to all the digital information that existed before 2003. By 2025 the total created *annually* will be over 160 zettabytes (or 160 trillion gigabytes). Global silicon reserves provide only a tiny fraction of what we would need to permanently store all the data we produce. But a single gram of DNA can store 455 trillion gigabytes.

Pak Chung Wong, a researcher at the Pacific Northwest National Laboratory, began to think about the problem of data storage in the early 2000s. The problem wasn't just the enormous quantity of information, but that the means of storage were insecure. For thousands of years, we have overcome entropy by leaving deliberate traces of what we know. Writing has allowed us to beat time and speak to the future, but so far, the materials available to us have been fragile: 'Bones and stone erode', Wong wrote, 'paper disintegrates, and electronic memory degrades.' Life, on the other hand, is in its essence defined by the safe transmission of information through time.

Wong imagined that he could exploit this very basic fact of life by converting binary information into the four nucleobases of DNA (with A for 00, G for 01, C for 10 and T for 11) and safely enclosing it in the genome of living matter, where, in theory, it would be retrievable for ever. In 2003, to test his idea, he encoded the lyrics to the Disney song 'It's a Small World After All' in the genomes of two bacteria – *Escherichia coli* (*E. coli*) and *Deinococcus radiodurans* (*D. radiodurans*).

Since Wong's experiment, others have radically increased the quantity of data stored in bacterial DNA. Researchers have encoded all kinds of information, from the full sequence of Shakespeare's sonnets to a PDF of Francis Crick and James Watson's Nobel Prize-winning paper on the structure of DNA; each time, however, they found that DNA could be sequenced only once to retrieve the information before it was destroyed. They could read the data, albeit with a few stray errors, but doing so effectively erased it. Storing data this way would necessitate having unique copies for every time the information was accessed, like keeping dozens of copies of the favourite book you read every year, or an ocean full of bottles each bearing the same message.

In 1953 – the same year that Crick and Watson published their paper and provided the new understanding of life that would make it possible to extend the central dogma – Philip K. Dick's short story 'The Preserving Machine' appeared in *The Magazine of Fantasy and Science Fiction*. In Dick's tale, a scientist called Doc Labyrinth becomes worried about the decline of civilization. He frets that the achievements of the present will be lost just as the wonders of the ancient world fell into darkness. The loss of music troubles him the most, so he devises a 'preserving machine' that allows him to transform musical scores into living things. Fed into Doc Labyrinth's machine, two popular songs emerge as scampering mice; a Mozart symphony becomes a small bird with a peacock's plume; a Stravinsky score produces a curious bird made up of odd fragments. Other composers' work is transformed into insects – a Beethoven beetle and a centipede-like Brahms creature – or novel species like the deeply coloured and vola-tile Wagner animal. Each has its own temperament, some doc-ile and some turbulent. Labyrinth releases his creations into the woods behind his house, but they swiftly become feral, feeding on one another and filling the night with screams. Disturbed by the turn of events, the doctor catches one of the Bach bugs and feeds it back into his preserving machine. The music that emerges is unrecognizable, an unearthly sound unlike anything he has heard before.

In looking for future fossils, I have encountered some start-ling means of recording our traces, from Renata Ferrari and Will Figueira's digital map of the Great Barrier Reef to the immense ice sheets of Greenland and Antarctica. The 'urban stratum' itself, the thin layer that will be all that is left of our greatest cities hundreds of millions of years from now, is a kind of physical archive foretold, foreshadowed by the landfill sites that accept those cities' rubbish today. The internet has allowed us to create the most detailed portrait of a civiliza-tion ever made, recording billions of interactions and images every day, stored away in large, hot data centres. But whether these archives were made intentionally or otherwise, none of them have aspired to write their record in the very stuff of life. DNA storage offers the prospect that our stories will be laid down in living, self-replicating archives.

The idea left me with a creeping sense of unease. There's an acute irony in the fact that scientists are contemplating DNA storage while global biodiversity is crashing. Treating other forms of life simply as resources is a large part of the mess we find ourselves in, but when our care ought to be directed towards fostering a viable future for all living things, we have begun to look to life itself to secure our own stories. But I also found the prospect of putting the things we value most in the care of microbes unsettling. I wondered if, when retrieving information comes to involve a visit to the laboratory rather than the library and what we want to recall is stored in vials rather than volumes, we would still welcome what emerges as ours. If a warp enters the weave, introducing a strain of something alien, what then? We may store the complete works of Shakespeare as a synthesized microbial memory, but if they return like the versions of great works in Borges's library, which differ from the original even if only in a single error, we may have to think again about what we're actually reading, and what kind of information – what *life* – it represents. Doc Labyrinth chooses to preserve music that represents the pinnacle of human achievement, but the sounds that he recovers from his menagerie have become deeply inhuman, twisted into monstrosity during their sojourn in the organic archive.

Perhaps, though, such concerns are misplaced. In 2017, researchers at the University of Washington successfully encoded recordings of Deep Purple's 'Smoke on the Water' and Miles Davis's 'Tutu' in DNA. When the information was retrieved, both songs played perfectly.

IN 2003, the experimental poet Christian Bök read Wong's paper and began to wonder if he could use bacterial DNA not just as a surface to write upon, but also as a writing partner. Since then, at a cost of hundreds of thousands of dollars, he has been trying to turn *D. radiodurans* into a machine not to preserve information but to write an eternal poem.

D. radiodurans was discovered in 1956, living in a tin of corned beef. Scientists in Oregon were experimenting with the preservative potential of gamma radiation when they found a bacterium they couldn't sterilize. Its name means 'the dire seed, immune to radiation': it can survive extreme desiccation and one thousand times the dose of gamma radiation lethal to humans. At three thousand times the exposure that would kill us, the bacterium is weakened but still viable. In 2002, NASA exposed a sample to ultraviolet solar radiation three hundred kilometres above the Earth for over six minutes, but *D. radiodurans* returned to Earth unharmed.

Its extraordinary resilience is simply down to its shape. The bacterium consists of four cells arranged in a snug ring and looks a little like a hot cross bun. Radiation damage breaks DNA apart, but *D. radiodurans*'s tightly bound shape means that the DNA is kept close together even when it is broken, allowing the microbe to rapidly repair itself. To the best of all knowledge, it is unkillable. A poem written in the genome of *D. radiodurans*, Christian Bök imagined, would 'persist on the planet until the sun itself explodes'.

A few days after I visited the Ranger uranium mine, I met up with Christian in Darwin, where he teaches creative writing. Just a few hours before I was due to fly back to Sydney, I found myself hurrying through the sauna-like streets of suburban Darwin to a seafront café on the rim of the continent to meet a poet endeavouring to write a poem that will live for ever.

The sea had turned lavender under the intense sun, and I was sweating and flustered when I arrived. Christian, by contrast, was relaxed and easy. I ordered a cold juice while he drank an espresso and a glass of sparkling wine and explained his process. Whereas researchers like Wong look for ways to preserve encoded information unchanged, Christian's aim was to encourage the bacterium to alter what he wrote. The core of the project is a pair of sonnets called the Xenotext; individually, the sonnets are named 'Orpheus' and 'Eurydice'. The first step was to find a cipher that would allow him to write a poem that, when read through the cipher, would translate the first poem into a second, entirely different one. 'I discovered the poems rather than wrote them', he said. He wrote a computer program that searched nearly eight trillion possibilities – most of which simply produced nonsense – before arriving at a viable cipher (called ANY-THE 112) that would transform one readable poem into another: so 'any style of life / is prim' – the first lines of 'Orpheus' – becomes 'the faery is rosy / of glow', the corresponding first lines of 'Eurydice'. Having done this, he selected twenty-six codons (sets of three nucleobases, which act as instructions for translating DNA into RNA) and assigned each to a letter of the alphabet. When the sonnet 'Orpheus' is translated into synthesized codons and introduced to the bacterium, *D. radiodurans* should 'read' the poem as a set of instructions for making a protein that, when read back through the cipher, emerges as 'Eurydice'. Like Orpheus passing through the underworld in search of his lost love, Christian's poem enters the unkillable bacterium in search of its partner.

In Ovid's poem, Orpheus is permitted to enter the underworld on the condition that he doesn't look at his wife until they are back on the surface. But at the last moment he is unable to help himself and has to watch her fall back into darkness. Orpheus' grief at losing Eurydice a second time overwhelms him, so much so that, even after death, his severed head continues to sing a lament as it floats on the waters of the Hebrus. His shade passes to the underworld again, where this time he is reunited for ever with his wife. 'There they stroll together, side by side', Ovid writes, 'or sometimes he leads the way and looks back, as he can do safely now, at his Eurydice.'

On New Year's Day 2019, 'Orpheus' and 'Eurydice' passed Ultima Thule, a belt of rubble left over from the formation of our solar system, six and a half billion kilometres from Earth, on board NASA's *New Horizons* spacecraft, and an extract from the sonnets now lies on the surface of Mars, included in the payload of the InSight lander. Christian has gone to extraordinary lengths to ensure the longevity of his poems, but if his experiment with microbial life is successful, the Xenotext will outlast every other future fossil. A million years from now, when the nuclear waste inside Onkalo is no more dangerous than a slab of butter, and cities like New Orleans and Shanghai are pressed flat beneath several kilometres of mud and clay, *D. radiodurans* will continue to write. When the only remaining traces of human presence on the planet's surface are the featureless, wind-eroded faces staring blindly from Mount Rushmore, it will be unmoved; until the dying sun absorbs Earth, 'Orpheus' and 'Eurydice' will persist in their duet.

For now, though, 'Eurydice' remains elusive. After fourteen years of experiments, Christian managed to verify that 'Orpheus' had been introduced into the bacterium, and that it was making a protein in response. But so far he had been unable to synthesize the protein and bring 'Eurydice' back to the light. *D. radiodurans* isn't cooperating. 'It's like I'm negotiating with this organism', Christian said. 'It doesn't matter what we try. Curse it, break it apart – it refuses to engage.'

He smiled ruefully. 'It's like trying to appease a little god.'

CODA

SEEING THE NEW WORLD

The sky was clear, just a few cottony cirrus clouds drifting across the face of the blue. It was March, and only a few weeks earlier Edinburgh had been smothered in thick snow, but today it felt warm enough to stuff my coat into my bag and let the breeze off the North Sea play through my thin sweater. The winter had lingered much longer than usual, weighing more heavily as each week passed without a change in the weather, but the milder air seemed to promise, finally, that a turn was on the way.

It was also the time of year when I take my students to Dunbar, to walk along the beach where the deep past meets the deep future. We had arrived by train that morning and were following the coast to the lighthouse, keeping to the pebble beach to avoid the golfers stalking the links. The tide was ebbing, and gulls dipped and wheeled over the water; perched on a rock farther out, where the tide had yet to recede, a cormorant was drying its wings. This stretch of coast joins a limestone pavement that juts out below the tideline, and when the water withdraws, it reveals a platform rich in fossil traces of ancient life. When we reached the lighthouse I knew we'd see the squat shape of the nuclear power station in the distance, and already the blank industrial buildings of the cement works were visible above the trees. I wanted my students to see how the traces cast in the limestone pavement three hundred million years ago also heralded a future of human legacies, so my gaze was cast down at what the slick rocks beneath our feet might reveal.

We were examining a block of sandstone undulating with the petrified marks of an ancient vanished tide when I saw something out of the corner of my eye. At first I thought it was part of the heap of seaweed dumped at the wrack line. But it was more like a sprig of heather, transplanted from a distant hillside. Thousands of pale orange and turquoise fronds, each thread-thin, curled and waved like antennae across the top of a bowl of piebald puddingstone large enough to fill both my hands. The top was stuck with grit, which gave it a sandy, sun-bleached look from above. It seemed impossible to tell if it was vegetable or mineral. Stone merged into something that looked organic and back into stone again.

In Paul Valéry's *Eupalinos*, the ghost of Socrates describes walking along 'an endless shore' long ago in his youth. The beach is full of things that the sea has rejected but the land has refused to reclaim – the charred timbers of shipwrecks and the collapsed forms of sea monsters. He comes across an object that startles him: something white, hard and smooth, but also indistinct, and he puzzles over what it could be and where it may have come from. Maybe it is a fish bone worn smooth, or a piece of carved ivory, the image of a god lost in the shipwreck. Or perhaps, he reasons, it was 'but the fruit of an infinite lapse of time'.

We peered closer at the strange object we'd discovered, prodding the branches. They were stiff and springy in a way that felt synthetic, but the shape was unlike anything we could categorize. It was as heavy as a bowling ball when I lifted it up.

As I peeled back some of the firm, strangely tentacular fronds, we saw that they hid a thick tube, slightly ridged and mostly covered with sand and grit. This wasn't a plant, I realized; it was a length of fishing rope that had become fused with the rock. The rope had frayed crazily at each end, unravelling its thousands of fibres, which now hung, shaggy with sand, like some kind of alien foliage. Our odd object was a plastiglomerate, a new kind of Anthropocene stone usually formed when beach fires melt together plastic debris with rocks and sediment grains. The first plastiglomerates were found in Hawaii in 2006, but they've since turned up on beaches worldwide, like harbingers of a strange new world.

Future fossils pose the peculiar challenge of learning to see a change that is both promised and already arrived. So often, on my journey through the deep future, I had seen the brief flash of *enargeia* break in on the present: in the glint of sunlight on the towers of Pudong and the shining sails of the Queensferry Crossing; the glistening ice core I'd held and the cream-warm glow of the coral cores printed with thousands of gaping Gorgon mouths. The bright lids of the copper deposition canisters, buried in the dark of Onkalo, and the shadow of extinction racing around the planet, threatening to engulf the light of life. The slick mud that had received the Happisburgh footprints so long ago and the gleaming black footprint in Edward

Burtynsky's photograph. Although its surfaces were dull, like the mysterious white object discovered by Socrates, our weird plastiglomerate plant-rock blazed with a similar, strange light.

'If a New World were discovered now', Italo Calvino once asked, 'would we be able to see it?' The signs of this new reality are all around us, in landscapes and objects that shimmer and shift at the edge of vision. What seems most transient conceals a potential to endure that sets the mind spinning. Some of our marks on deep time are inevitable, already guaranteed by the scale of the cities and roads we have built and the durability of the materials we have devised; the extent of others remains contingent on the choices we have yet to make, like the emptying of ecosystems or the catastrophic thaw of frozen methane reserves. But the importance of the task – to see here and now changes that will unfold over many lifetimes, and to feel our closeness to those lives as an acute, intimate responsibility – can't be overstated. Future fossils show us that we aren't obligated only to the generations that will directly follow ours, the children of our children's children, but to humans who are separated from us by hundreds, even thousands of generations. These are people whose languages and cultures will be wholly alien to what we know or can imagine, but who may have to live in a world still warped by our decisions, made millennia before their birth. The better we learn to see the new world that is promised by our inaction, the better, I believe, we will be able to imagine an alternative – for ourselves and for those who will follow us.

Still, it is far from easy. We think we know what to expect from the world we live in and miss the opportunity to see things not only for what they are but also for what they are becoming. New worlds open up every day, Calvino observed, and we fail to notice them. In the rush of everyday life we miss the subtle shift; through habit, we see the present by the light of the past. The challenge is to learn instead to examine our present, and ourselves, by the eerie light cast by the onrushing future.

Socrates' shade recalls how disturbed he was by his perplexing object. 'I could not determine', he recounts, 'whether this singular object were the work of life, or of art, or rather of time'. Suddenly, in frustration, he pitches it back into the sea, but what he has seen stays with him, leaving him altered in ways he cannot fully describe. Even as a ghost, he remembers its troubling whiteness. I knew, standing on the familiar beach, that our weird discovery would remain with me for ever. This was the new world, here, in my hands.

We spent a few more minutes examining our curious find, turning it this way and that like prospectors searching for a vein of gold. But it was too heavy to carry away with us, and so I left it perched on a rock above the tideline. The sky began to fill with grey as we carried on up the beach towards the lighthouse, and our steps left no prints on the pebble beach.

SELECTED BIBLIOGRAPHY

INTRODUCTION: TRACES OF A HAUNTED FUTURE

Anthony Andrady, *Plastics and Environmental Sustainability* (John Wiley, 2015); David Archer, 'Fate of Fossil Fuel CO_2 in Geologic Time', *Journal of Geophysical Research* 110 (2005); David Archer and Victor Brovkin, 'The Millennial Atmospheric Lifetime of Anthropogenic CO_2', *Climate Change* 90 (2008); Aristotle, *Rhetoric*, trans. Lane Cooper (Appleton-Crofts, 1932); Thomas Carlyle, 'Boswell's Life of Johnson', *Fraser's Magazine* 5, no. 28 (May 1832); Damian Carrington, 'How the Domestic Chicken Rose to Define the Anthropocene', *Guardian*, 31 August 2016; John Stewart Collis, *The Worm Forgives the Plough* (Penguin, 1975); Daniel Defoe, *Robinson Crusoe* (Oxford University Press, 2007); T. S. Eliot, *Complete Poems: 1909–1962* (Faber, 2009); Owen Gaffney and Will Steffen, 'The Anthropocene Equation', *Anthropocene Review* 4, no. 1 (2017); William Grimes, 'Seeking the Truth in Refuse', *New York Times*, 13 August 13 1992; Roger LeB. Hooke, 'On the History of Humans as Geomorphic Agents', *Geology* 28, no. 9 (2000); Richard Irvine, 'The Happisburgh Footprints in Time', *Anthropology Today* 30, no. 2 (2014); Adam Nicolson, *The Seabird's Cry* (William Collins, 2017); Alice Oswald, *Memorial* (Faber, 2011); Stephanie Pappas, 'Human Ancestor 'Family' May Not Have Been Related', Live Science, 4 November 2011, https://www.livescience.com/16894-human-ancestor-laetoli-footprints-family.html; Heinrich Plett, *Enargeia in Classical Antiquity and the Early Modern Age*

(Brill, 2012); Percy Bysshe Shelley, *The Major Works* (Oxford University Press, 2003); Robert Louis Stevenson, 'A Gossip on Romance', *Longman's Magazine* 1, no. 1 (November 1882); James Temperton, 'Inside Sellafield: How the UK's Most Dangerous Nuclear Site Is Cleaning Up Its Act', *Wired*, 17 September 2016, https://www.wired.co.uk/article/inside -sellafield-nuclear-waste-decommissioning; Alfred, Lord Tennyson, *In Memoriam* (W. W. Norton, 2004); Bruce Wilkinson, 'Humans as Geo-logic Agents: A Deep-Time Perspective', *Geology* 33, no. 3 (2005); Jan Zalasiewicz and Katie Peek, 'A History in Layers', *Scientific American* 315, no. 3 (2016).

1. THE INSATIABLE ROAD

J. G. Ballard, *Extreme Metaphors: Collected Interviews* (Fourth Estate, 2014); Vince Beiser, 'The Deadly Global War for Sand', *Wired*, 26 March 2015, https://www.wired.com/2015/03/illegal-sand-mining; A. G. Brown et al., 'The Anthropocene: Is There a Geomorphological Case?', *Earth Surface Processes and Landforms* 38, no. 4 (2013); Edward Burtynsky, *Manufactured Landscapes: The Photography of Edward Burtynsky* (National Gallery of Canada, 2003); Edward Burtynsky, *Quarries* (Steidl, 2007); Edward Burtynsky, *Oil* (Steidl/Corcoran, 2009); Bruce Chatwin, *In Patagonia* (Picador, 1977); Hart Crane, *The Complete Poems of Hart Crane* (Liveright, 2001); Joan Didion, *The White Album* (Farrar, Straus and Giroux, 1979); Ralph Waldo Emerson, *Emerson's Prose and Poetry* (W. W. Norton, 2001); Roy Fisher, *The Long and the Short of It: Poems 1955–2010* (Bloodaxe, 2012); Seamus Heaney, *Station Island* (Faber, 1984); Roger LeB. Hooke, 'On the Efficacy of Humans as Geomorphic Agents', *GSA Today* 4, no. 9 (1994); Ryszard Kapuściński, *Shah of Shahs*, trans. William R. Brand and Katarzyna Mroczkowska-Brand (Penguin, 2006); KCGM, 'Mineral Processing', http://www.superpit.com.au/about /mineral-processing/; Jack Kerouac, *On the Road* (Penguin, 1991); Barry Lopez, *Arctic Dreams* (Picador, 1986); Michael Mitchell, 'More Urgent Than Beauty', in Edward Burtynsky, *Quarries* (Steidl, 2007); Ben Okri, *The Famished Road* (Vintage, 1992); David Owen, 'The World Is Running Out of Sand', *New Yorker*, 22 May 22 2017; Val Plumwood, 'Shadow Places and the Politics of Dwelling', *Australian Humanities Review* 44 (March 2008); E. Ramirez-Llodra, 'Man and the Last Great Wilderness: Human Impact on the Deep Sea', *PLoS One* 6, no. 8 (2011); Neil L. Rose, 'Spheroidal Carbonaceous Fly Ash Particles Provide a Globally

Synchronous Stratigraphic Marker for the Anthropocene', *Environmental Science and Technology* 49, no. 7 (2015); Wolfgang Schivelbusch, *The Railway Journey: The Industrialization and Perception of Time and Space* (University of California Press, 1977); Autumn Spanne, 'We're Running Out of Sand', *Mental Floss*, 21 June 2015, https://www.mentalfloss.com /article/65341/were-running-out-sand; Christopher Simon Sykes, *Hockney: A Pilgrim's Progress* (Century, 2011); James P. M. Syvitski and Albert J. Kettner, 'Sediment Flux and the Anthropocene', *Philosophical Transactions of the Royal Society A* 369 (2011); Edward Thomas, *The Icknield Way* (Wildwood House, 1980); Michael Torosian, 'The Essential Element: An Interview with Edward Burtynsky', in *Manufactured Landscapes: The Photography of Edward Burtynsky*, ed. Lori Pauli (National Gallery of Canada, 2003); Gaia Vince, *Adventures in the Anthropocene* (Chatto and Windus, 2014); Jan Zalasiewicz, *The Earth After Us* (Oxford University Press, 2008); Jan Zalasiewicz et al., 'Human Bioturbation, and the Subterranean Landscapes of the Anthropocene', *Anthropocene* 6 (2014); Jan Zalasiewicz et al., 'Petrifying Earth Process: The Stratigraphic Imprint of Key Earth System Parametres in the Anthropocene', *Theory, Culture, and Society* 34, nos. 2–3 (2017).

2. THIN CITIES

Peter Ackroyd, *Venice: Pure City* (Vintage, 2010); J. G. Ballard, *Extreme Metaphors: Collected Interviews* (Fourth Estate, 2014); J. G. Ballard, *Miracles of Life* (Fourth Estate, 2014); J. G. Ballard, *The Drowned World* (Fourth Estate, 2012); Walter Benjamin, *The Arcades Project*, trans. Howard Eiland and Kevin McLaughlin (Harvard University Press, 1999); Walter Benjamin, *Illuminations*, trans. Harry Zohn (Fontana/Collins, 1979); Daniel Brook, *A History of Future Cities* (Norton, 2013); Italo Calvino, *Hermit in Paris*, trans. Martin McLaughlin (Jonathan Cape, 2003); Italo Calvino, *Invisible Cities*, trans. William Weaver (Picador, 1979); Richard Campanella, 'How Humans Sank New Orleans', *Atlantic*, 6 February 2018; 'China Is Trying to Turn Itself into a Country of 19 Super-Regions', *Economist*, 23 June 2018; J. A. Church et al., 'Sea Level Change', in *Climate Change 2013: The Physical Science Basis. Contribution of Working Group I to the Fifth Assessment Report of the Intergovernmental Panel on Climate Change*, eds. T. F. Stocker et al. (Cambridge University Press, 2013); Lisa Cox, 'Cavity Two-Thirds the Size of Manhattan Discovered Under Antarctic Glacier', *Guardian*, 6 February 2019; Orlando

Croft, 'China's Atlantis: How Shanghai Is Slowly Sinking Under the Weight of Its Tallest Towers', *IB Times*, 9 January 2017; *The Epic of Gilgamesh*, trans. Andrew George (Penguin, 2003); Jeff Goodell, *The Waters Will Come* (Black, 2018); O. Hoegh-Guldberg et al., 'Impacts of 1.5°C Global Warming on Natural and Human Systems', in *Global Warming of 1.5°C*, ed. V. P. Masson-Delmotte et al. (World Meteorological Organization, 2018); Hurricane Katrina External Review Panel, *The New Orleans Hurricane Protection System: What Went Wrong and Why* (ASCE Press, 2007); IPCC, 'Summary for Policymakers', in *Global Warming of 1.5°C*, ed. V. P. Masson-Delmotte et al. (World Meteorological Organization, 2018); Frederic Lane, *Venice: A Maritime History* (Johns Hopkins University Press, 1973); Leo Ou-Fan Lee, *Shanghai Modern: The Flowering of New Urban Culture in China, 1930–1945* (Harvard University Press, 1999); Coco Lui, 'Shanghai Struggles to Save Itself from the Sea', *New York Times*, 27 September 2011; Hugh MacDiarmid, *Selected Poetry* (Carcanet, 2004); Joe McDonald, 'Shanghai Is Sinking', ABC News, 28 July 2000; Robert I. McDonald et al., 'Urbanization and Global Trends in Biodiversity and Ecosystem Services', in *Urbanization, Biodiversity and Ecosystem Services: Challenges and Opportunities*, ed. Thomas Elmqvist et al. (Springer, 2013); Gordon McGranahan et al., 'Low Coastal Zone Settlements', *Tiempo* 59 (2006), https://sedac.ciesin.columbia.edu/downloads/docs/lecz/coastal_tiempo.pdf; Olga Mecking, 'Are the Floating Houses of the Netherlands a Solution Against Rising Seas?', *Pacific Standard*, 21 August 2017, https://www.psmag.com/environment/are-the-floating-houses-of-the-netherlands-a-solution-against-the-rising-seas; P. Milillo et al., 'Heterogeneous Retreat and Ice Melt of Thwaites Glacier, West Antarctica', *Science Advances* 5, no. 1 (2019); Edward Muir, *Civic Ritual in Renaissance Venice* (Princeton University Press, 1981); Jaap H. Nienhuis et al., 'A New Subsidence Map for Coastal Louisiana', *GSA Today* 27, no. 9 (2017); John Ruskin, *Stones of Venice* (Dana Estes, 1851); W. G. Sebald, *Vertigo*, trans. Michael Hulse (Harvill Press, 1990); Mu Shying, *China's Lost Modernist*, trans. Andrew David Field (Hong Kong University Press, 2014); J. D. Stanford et al., 'Sea-Level Probability for the Last Deglaciation: A Statistical Analysis of Far-Field Records', *Global and Planetary Change* 79, nos. 3–4 (2011); UN-Habitat, *Urbanisation and Development: Emerging Futures World Cities Report* (2016); Union Internationale des Transports Publics, *World Metro Figures: Statistics Brief*, October 2015, https://www.uitp.org/sites/default/files/cck-focus-papers-files/UITP

-Statistic%20Brief-Metro-A4-WEB_0.pdf; US Department of Housing and Urban Development, *The Big 'U': Rebuild by Design*, http://www. rebuildbydesign.org; *What the World Would Look Like If All the Ice Melted*, September 2013, https://www.nationalgeographic.com; P. P. Wong et al., 'Coastal Systems and Low-Lying Areas', in *Climate Change 2014: Impacts, Adaptation, and Vulnerability. Part A: Global and Sectorial Aspects. Contribution of Working Group II to the Fifth Assessment Report of the Intergovernmental Panel on Climate Change*, ed. C. B. Field et al. (Cambridge University Press, 2014); *World Ocean Review 5: Coasts – A Vital Habitat Under Pressure* (Maribus, 2017), https://www.worldoceanreview.com/en /wor-5/; Qiu Xiaolong, *A Case of Two Cities* (Hodder and Stoughton, 2006); Jan Zalasiewicz, *The Earth After Us* (Oxford University Press, 2008).

3. THE BOTTLE AS HERO

Anthony Andrady, *Plastics and Environmental Sustainability* (Wiley, 2015); David K. A. Barnes et al., 'Accumulation and Fragmentation of Plastic Debris in Global Environments', *Philosophical Transactions of the Royal Society B* 364 (2009); Roland Barthes, *Mythologies*, trans. Annette Lavers (Paladin, 1987); Bernadette Bensaude-Vincent, 'Plastics, Materials, and Dreams of Dematerialization', in *Accumulation: The Material Politics of Plastic*, ed. Jennifer Gabrys et al. (Routledge, 2013); C. M. Boerger et al., 'Plastic Ingestion by Planktivorous Fishes in the North Pacific Central Gyre', *Marine Pollution Bulletin* 60, no. 12 (2010); Mark A. Browne et al., 'Spatial Patterns of Plastic Debris Along Estuarine Shorelines', *Environmental Science and Technology* 44, no. 9 (2010); Matthew Cole et al., 'Microplastic Ingestion by Zooplankton', *Environmental Science and Technology* 47, no. 12 (2013); Patricia L. Corcoran et al., 'An Anthropogenic Marker Horizon in the Future Rock Record', *GSA Today* 24, no. 6 (2014); Patricia L. Corcoran et al., 'Hidden Plastics of Lake Ontario', *Environmental Pollution* 204 (2015); Marcus Eriksen et al., 'Plastic Pollution in the World's Oceans: More Than 5 Trillion Plastic Pieces Weighing Over 250,000 Tons Afloat at Sea', *PLoS One* 9, no. 12 (2014); Jan A. Franeker and Kara Lavender Law, 'Seabirds, Gyres and Global Trends in Plastic Pollution', *Environmental Pollution* 203 (2015); Roland Geyer et al., 'Production, Use, and Fate of All Plastics Ever Made', *Science Advances* 3, no. 7 (2017); William Golding,

The Inheritors (Faber, 1955); Murray Gregory, 'Environmental Implications of Plastic Debris in Marine Settings', *Philosophical Transactions of the Royal Society B* 364 (2009); E. A. Howell et al., 'On North Pacific Circulation and Associated Marine Debris Concentration', *Marine Pollution Bulletin* 65, nos 1–3 (2012); Juliana A. Ivar do Sul and Monica F. Costa, 'The Present and Future of Microplastic Pollution in the Marine Environment', *Environmental Pollution* 185 (2014); Mark Jackson, 'Plastic Islands and Processual Grounds: Ethics, Ontology, and the Matter of Decay', *Cultural Geographies* 20, no. 2 (2012); Sarah Laskow, 'How the Plastic Bag Became So Popular', *Atlantic*, 10 October 2014; Bruno Latour, *Pandora's Hope: Essays on the Reality of Science Studies* (Harvard University Press, 1999); L. C.-M. Lebreton et al., 'Numerical Modelling of Floating Debris in the World's Oceans', *Marine Pollution Bulletin* 64, no. 3 (2012); Ursula K. Le Guin, 'The Carrier Bag Theory of Fiction', in *Women of Vision*, ed. Denise DuPont (St. Martin's Press, 1988); Jeffrey Meikle, *American Plastic: A Cultural History* (Rutgers University Press, 1997); Christopher K. Pham et al., 'Marine Litter Distribution and Density in European Seas, from the Shelves to the Basins', *PLoS One* 9, no. 4 (2014); William G. Pichel et al., 'Marine Debris Collects Within the North Pacific Subtropical Convergence Zone', *Marine Pollution Bulletin* 54, no. 8 (2007); Peter G. Ryan et al., 'Monitoring the Abundance of Plastic Debris in the Marine Environment', *Philosophical Transactions of the Royal Society B* 364 (2009); Xavier Tubau et al., 'Marine Litter on the Floor of Deep Submarine Canyons of the Northwestern Mediterranean Sea', *Progress in Oceanography* 134 (2015); Lisbeth Van Cauwenberghe et al., 'Microplastic Pollution in Deep-Sea Sediments', *Environmental Pollution* 182 (2013); Lucy C. Woodall et al., 'The Deep Sea Is a Major Sink for Microplastic Debris', *Royal Society Open Science* 1, no. 4 (2014); R. Yamashita and A. Tanimura, 'Floating Plastic in the Kuroshio Current Area, Western North Pacific Ocean', *Marine Pollution Bulletin* 54, no. 4 (2007); Jan Zalasiewicz et al., 'The Geological Cycle of Plastics and Their Use as a Stratigraphic Indicator of the Anthropocene', *Anthropocene* 13 (2016); Eric Zettler et al., 'Life in the 'Plastisphere': Microbial Communities on Plastic Marine Debris', *Environmental Science and Technology* 47, no. 13 (2013).

4. THE LIBRARY OF BABEL

Richard Alley, *The Two-Mile Time Machine: Ice Cores, Abrupt Climate Change, and Our Future* (Princeton University Press, 2000); Matthew Amesbury et al., 'Widespread Biological Response to Rapid Warming on the Antarctic Peninsula', *Current Biology* 27, no. 11 (2017); Alessandro Antonello, 'Engaging and Narrating the Antarctic Ice Sheet', *Environmental History* 22, no. 1 (2017); Alessandro Antonello and Mark Carey, 'Ice Cores and the Temporalities of the Global Environment', *Environmental Humanities* 9, no. 2 (2007); Jonathan Bate, *The Song of the Earth* (Picador, 2000); Tom Bawden, 'Global Warming: Data Centres to Consume Three Times as Much Energy in Next Decade', *Independent*, 23 January 2016; Jorge Luis Borges, *Labyrinths*, trans. James E. Irby (Penguin, 2000); James W. P. Campbell, *The Library: A World History* (Thames and Hudson, 2013); Mark Carey, 'The History of Ice: How Glaciers Became an Endangered Species', *Environmental History* 12, no. 3 (2007); Damian Carrington, 'A Third of Himalayan Ice Cap Doomed, Finds Report', *Guardian*, 4 February 2019; Joseph Cheek, 'What Ice Cores from Law Dome Can Tell Us About Past and Current Climates', 12 August 2011, https://www.sciencepoles.org/interview/what-ice-cores-from-law-dome-can-tell-us-about-past-and-current-climates; William Colgan et al., 'The Abandoned Ice Sheet Base at Camp Century, Greenland, in a Warming Climate', *Geophysical Research Letters* 43, no. 15 (2016); DOMO, 'Data Never Sleeps 6.0', https://www.domo.com/learn/data-never-sleeps-6; Aant Elzinga, 'Some Aspects in the History of Ice Core Drilling and Science from IGY to EPIPCA', in *National and Trans-National Agendas in Antarctic Research from the 1950s and Beyond*, ed. C. Lüdecke (Byrd Polar and Climate Research Centre, Ohio State University, 2013); Michel Foucault, 'Of Other Spaces', trans. Jay Miskoweic, *Diacritics* 16, no. 1 (1986); Gavin Francis, *Empire Antarctica: Ice, Silence and Emperor Penguins* (Chatto and Windus, 2012); A. Ganopolski et al., 'Critical Insolation-CO_2 Relations for Diagnosing Past and Future Glacial Inception', *Nature* 529 (2016); Tom Griffiths, 'Introduction: Listening to Antarctica', in *Antarctica: Music, Sounds and Cultural Connections*, ed. Bernadette Hince et al. (Australian National University Press, 2015); O. Hoegh-Guldberg et al., 'Impacts of 1.5°C Global Warming on Natural and Human Systems', in *Global Warming of 1.5°C*, ed. V. P. Masson-Delmotte et al. (World Meteorological Organization, 2018); Adrian Howkins, 'Melting Empires? Climate Change and Politics in Antarctica Since the International Geophysical Year', *Osiris* 26, no. 1 (2011); Alexander Koch, 'Earth System Impacts of the European Arrival and Great Dying in the Americas After 1492', *Quaternary Science Reviews* 207 (2019); Tété-Michel Kpomassie, *An African in Greenland*, trans. James Kirkup (New York Review Books, 2001); Chester C. Langway, Jr., *The History of Early Polar Ice Cores* (Engineer Research and Development Centre, 2008); Kurd Lasswitz, 'The Universal Library', in *Fantasia Mathematica* (Simon & Schuster, 1958); Jasmine R. Lee et al., 'Climate Change Drives Expansion of Antarctic Ice-Free Habitat', *Nature* 547 (2017); Matthieu Legendre et al., 'In-Depth Study of *Mollivirus sibericum*, a New 30,000-y-Old Giant Virus Infecting *Acanthamoeba*', *Proceedings of the National Academy of Sciences* 112, no. 38 (2015); Alec Luhn, 'Anthrax Outbreak Triggered by Climate Change Kills Boy in Arctic Circle', *Guardian*, 1 August 2016; D. R. MacAyeal, 'Seismology Gets Under the Skin of the Antarctic Ice Sheet', *Geophysical Research Letters* 45, no. 20 (2018); Janet Martin-Nielsen, '"The Deepest and Most Rewarding Hole Ever Drilled": Ice Cores and the Cold War in Greenland', *Annals of Science* 70, no. 1 (2013); Oliver Milman, 'US Glacier National Park Losing Its Glaciers with Just 26 of 150 Left', *Guardian*, 11 May 2017; Jing Ming et al., 'Widespread Albedo Decreasing and Induced Melting of Himalayan Snow and Ice in the Early 21st Century', *PLoS One* 10, no. 6 (2015); John Muir, *John Muir: His Life and Letters and Other Writings*, ed. Terry Gifford (Mountaineering Books, 1996); John Muir, 'Yosemite Glaciers', *New-York Tribune*, 5 December 1871; Kristian H. Nielsen et al., 'City Under the Ice: The Closed World of Camp Century in Cold War Culture', *Science as Culture* 23, no. 4 (2014); Rachel Obbard et al., 'Global Warming Releases Microplastic Legacy Frozen in Arctic Sea Ice', *Earth's Future* 2, no. 6 (2014); Alvin Powell, 'Study of 14th-Century Plague Challenges Assumptions on "Natural" Lead Levels', Phys.org, 31 May 2017, https://www.phys.org/news/2017-05-14th-century-plague-assumptions-natural.html; Project Ice Memory, https://fondation.univ-grenoble-alpes.fr; Radicati Group, Inc., *Email Statistics Report, 2017–2021*, February 2017, https://www.radicati.com/wp/wp-content/uploads/2017/01/Email-Statistics-Report-2017-2021-Executive-Summary.pdf; Arundhati Roy, 'What Have We Done to Democracy? Of Nearsighted Progress, Feral Howls, Consensus, Chaos, and a New Cold War in Kashmir', *TomDispatch*, 27 September 2009, http://www.tomdispatch.com/blog/175125/tomgram%3A_arundhati_roy%2C_is_democracy_melting; William Ruddiman, 'The

Anthropogenic Greenhouse Era Began Thousands of Years Ago', *Climate Change* 61, no. 3 (2003); William Ruddiman, 'How Did Humans First Alter Global Climate?', *Scientific American* 292, no. 3 (March 2005); Ted Schuur, 'The Permafrost Prediction', *Scientific American* 315, no. 6 (2016); Yun Lee Too, *The Idea of the Library in the Ancient World* (Oxford University Press, 2010); Peter Wadhams, *A Farewell to Ice* (Allen Lane, 2016); Walter Wager, *Camp Century: City Under the Ice* (Chilton Books, 1962); Eric N. Woolf, 'Ice Sheets and the Anthropocene', in *A Stratigraphic Basis for the Anthropocene*, ed. Colin Waters et al. (Geological Society of London, 2014).

The recording of Antarctic ice singing is available here: https://www.theguardian.com/global/video/2018/oct/18/researchers-capture-audio-of-antarctic-ice-singing-video.

5. MEDUSA'S GAZE

Theodor Adorno, *Prisms*, trans. Samuel and Shierry Weber (Massachusetts Institute of Technology Press, 1967); Joseph Banks, *The Endeavour Journal of Joseph Banks: The Australian Journey*, ed. Paul Brunton (Angus and Robertson, 1998); Tom Bawden, 'Caribbean Coral Reefs Are Declining at 'an Alarming' Rate', *Independent*, 2 July 2014; Thomas Browne, *Pseudodoxia Epidemica* 1 (Clarendon Press, 1981); Gilbert Camoin and Jody Webster, 'Coral Reefs and Sea-Level Change', *Developments in Marine Geology* 7 (2014); *The Correspondence of Charles Darwin*, ed. Frederick Burkhardt and Sydney Smith, vol. 1, *1821–1836* (Cambridge University Press, 1985); Adrian Desmond and James Moore, *Darwin* (Michael Joseph, 1991); Tim DeVries, 'Recent Increase in Oceanic Carbon Uptake Driven by Weaker Upper-Ocean Overturning', *Nature* 542 (2017); C. G. Ehrenberg, 'On the Nature and Formation of the Coral Islands and Coral Banks in the Red Sea', *Journal of the Bombay Branch of the Royal Asiatic Society* 1 (July 1841–July 1844); Great Barrier Reef Marine Park Authority, *Final Report: 2016 Coral Bleaching Event on the Great Barrier Reef* (GBRMPA, 2017); Jane Ellen Harrison, *Prolegomena to the Study of the Greek Religion* (Cambridge University Press, 2013); Stefan Helmreich, *Sounding the Limits of Life: Essays in the Anthropology of Biology and Beyond* (Princeton University Press, 2015); Terry Hughes et al., 'Ecological Memory Modifies the Cumulative Impact of Recurrent Climate Extremes', *Nature Climate Change* 9 (2019); Derek Jarman,

Chroma (Vintage, 2000); Elizabeth Kolbert, 'The Darkening Sea', *New Yorker*, 20 November 2006; Dan Lin and Kathy Jetñil-Kijiner, 'Dome Poem Part III: "Anointed" Final Poem and Video', 16 April 2018, https:// www.kathyjetnilkijiner.com/dome-poem-iii-anointed-final-poem-and -video/; Iain McCalman, *The Reef: A Passionate History* (Scribe, 2014); Mathelinda Nabugodi, 'Medusan Figures: Reading Percy Bysshe Shelley and Walter Benjamin', *MHRA Working Papers in the Humanities* 9 (2015); Patrick D. Nunn and Nicholas J. Reid, 'Aboriginal Memories of Inundation of the Australian Coast Dating from More Than 7,000 Years Ago', *Australian Geographer* 47, no. 1 (2016); Ovid, *Metamorphoses*, trans. Mary Innes (Penguin, 1955); Nicholas J. Reid et al., 'Indigenous Australian Stories and Sea-Level Change', in *Indigenous Languages and Their Value to the Community*, ed. Patrick Heinrich and Nicholas Ostler, Proceedings of the 18th Foundation for Endangered Languages Conference, Okinawa, Japan (2014); C. Sabine, 'Study Details Distribution, Impacts of Carbon Dioxide in the World Oceans', *NOAA Magazine*, 2014, http:// www.noaanews.noaa.gov; William Shakespeare, *The Tempest* (Bloomsbury, 2011); Derek Walcott, *Omeros* (Faber, 1990); Colin Woodroffe and Jody Webster, 'Coral Reefs and Sea Level Change', *Marine Geology* 352 (2014); Frances Yates, *The Art of Memory* (Routledge, 1966).

6. THE MOMENT UNDER THE MOMENT

Svetlana Alexievich, *Chernobyl Prayer*, trans. Anna Gunin and Arch Tait (Penguin, 2013); 'Australia's Uranium', World Nuclear Association, http://www.world-nuclear.org/information-library/country-profiles /countries-a-f/australia.aspx; David Bradley, *No Place to Hide* (University Press of New England, 1983); Julia Bryan-Wilson, 'Building a Marker of Nuclear Warning', in *Monuments and Memory, Made and Unmade*, ed. Robert S. Nelson and Margaret Olin (University of Chicago Press, 2003); Jane Dibblin, *Day of Two Suns: U.S. Nuclear Testing and the Pacific Islanders* (New Amsterdam, 1990); Herodotus, *The Histories*, trans. Aubrey de Sélincourt (Penguin, 1996); Russell Hoban, *The Moment Under the Moment* (Picador, 1992); International Atomic Energy Agency, *Estimation of Global Inventories of Radioactive Waste and Other Radioactive Materials*, IAEA-TECDOC-1591 (IAEA, 2008); Jawoyn Association, https://www .jawoyn.org.au; Barbara Rose Johnson, 'Nuclear Disaster: The Marshall Islands Experience and Lessons for a Post-Fukushima World', in *Global Ecologies and the Environmental Humanities: Postcolonial Approaches*, ed.

Anthony Carrigan et al. (Routledge, 2015); *The Kalevala*, trans. Keith Bosley (Oxford University Press, 2008); Martti Kalliala et al., *Solution 239–246 Finland: The Welfare Game* (Sternberg Press, 2011); Matti Kuusi et al., eds, *Finnish Folk Poetry – Epic: An Anthology in Finnish and English* (Finnish Literature Society, 1977); Joseph Masco, *The Nuclear Borderlands: The Manhattan Project in Post-Cold War New Mexico* (Princeton University Press, 2006); Andrew Moisey, 'Considering the Desire to Mark Our Buried Nuclear Waste: Into Eternity and the Waste Isolation Pilot Plant', *Qui Parle* 20, no. 2 (2012); 'The Nuclear Fuel Cycle', http://www.world-nuclear.org; Mark Pagel et al., 'Ultraconserved Words Point to Deep Language Ancestry Across Europe', *PNAS* 110, no. 21 (2013); *Permanent Markers Implementation Plan* (United States Department of Energy, 2004); Posiva, *Biosphere Assessment Report* (2009); Posiva, *Safety Case for the Disposal of Spent Nuclear Fuel at Onkalo – Complementary Considerations* (December 2012); Thomas Sebeok, *Communication Measures to Bridge Ten Millennia* (Office of Nuclear Waste Isolation, 1984); Sophocles, *Three Theban Plays*, trans. Robert Fagles (Penguin, 1984); Kathleen M. Trauth et al., *Expert Judgement on Markers to Deter Inadvertent Human Intrusion into the Waste Isolation Pilot Plant* (Sandia National Laboratories, 1993); Peter C. Van Wyck, *Signs of Danger: Waste, Trauma and Nuclear Threat* (University of Minnesota Press, 2005); Mark Willacy, 'A Poison in Our Island', ABC News, 26 November 2017, https://www.abc.net.au/news/2017-11-27/the-dome-runit-island-nuclear-test-leaking-due-to-climate-change/9161442; Alexis Wright, *Carpentaria* (Constable, 2006); Tom Zoellner, *Uranium: War, Energy, and the Rock That Shaped the World* (Viking, 2009).

7. WHERE THERE SHOULD BE SOMETHING, THERE IS NOTHING

Stacy Alaimo, 'Jellyfish Science, Jellyfish Aesthetics', in *Thinking with Water*, ed. Celia Chen et al. (McGill-Queens University Press, 2013); Baltic Marine Environment Protection Commission, *The State of the Baltic Sea* (2017); Lucas Brotz et al., 'Increasing Jellyfish Populations: Trends in Large Marine Ecosystems', *Hydrobiologia* 690, no. 1 (2012); J. W. Bull and M. Maron, 'How Humans Drive Speciation as Well as Extinction', *Proceedings of the Royal Society B* 283 (2016); Donald E. Canfield et al., 'The Evolution and Future of Earth's Nitrogen Cycle', *Science* 330, no. 6001 (2010); Robert Diaz and Rutger Rosenberg, 'Spreading Dead

Zones and Consequences for Marine Ecosystems', *Science* 321, no. 5891 (2008); Annie Dillard, *Teaching a Stone to Talk* (Canongate, 2017); T. S. Eliot, *Complete Poems: 1909–1962* (Faber, 2009); James J. Elser, 'A World Awash with Nitrogen', *Science* 334, no. 6062 (2011); Mark Fisher, *The Weird and the Eerie* (Repeater, 2016); Tim Flannery, 'They're Taking Over!', *New York Review of Books*, 26 September 2013; Shigehisa Furuya, 'World Worries as Jellyfish Swarms Swell', *Nikkei Asian Review*, 5 February 2015; Lisa-ann Gershwin, *Stung! On Jellyfish Blooms and the Future of the Ocean* (University of Chicago Press, 2013); Ernst Haeckel, *Art Forms in Nature* (Prestel, 1998); Lila M. Harper, '"The Starfish That Burns": Gendering the Jellyfish', in *Forces of Nature*, ed. Bernadette H. Hyner and Precious McKenzie Stearns (Cambridge Scholars, 2009); Intergovernmental Science-Policy Platform on Biodiversity and Ecosystem Services, *The Global Assessment Report on Biodiversity and Ecosystem Services*, E. S. Brondizio, ed. J. Settele, S. Díaz, and H. T. Ngo (IPBES Secretariat, 2019); Michael L. McKinney, 'How Do Rare Species Avoid Extinction? A Paleontological View', in *The Biology of Rarity*, ed. William E. Kumin and Kevin J. Gastin (Springer, 1997); Daniel Pauly, 'Anecdotes and the Shifting Baseline Syndrome of Fisheries', *Tree* 10 (1995); Jennifer E. Purcell, 'Jellyfish and Ctenophore Blooms Coincide with Human Proliferations and Environmental Perturbations', *Annual Review of Marine Science* 4 (2012); Robert J. Richards, *The Tragic Sense of Life: Ernst Haeckel and the Struggle Over Evolutionary Thought* (University of Chicago Press, 2006); Anthony J. Richardson et al., 'The Jellyfish Joyride: Causes, Consequences and Management Responses to a More Gelatinous Future', *Trends in Ecology and Evolution* 24, no. 6 (2009); Mark Schrope, 'Marine Ecology: Attack of the Blobs', *Nature* 482 (1 February 2012); Vaclav Smil, *The Earth's Biosphere* (MIT Press, 2002); Jean Sprackland, *Hard Water* (Jonathan Cape, 2003); Jens-Christian Svenning, 'Future Megaphones: A Historical Perspective on the Potential for a Wilder Anthropocene', *Arts of Living on a Damaged Planet*, ed. Anna Lowenhaupt Tsing et al. (University of Minnesota Press, 2017); Tomas Tranströmer, *New Collected Poems*, trans. Robin Fulton (Bloodaxe, 1997); John Vidal, 'UN Environment Programme: 200 Species Extinct Every Day', *HuffPost*, 18 August 2010, https://www.huffpost.com/entry/un-environment-programme_n_684562; Mary Wollstonecraft, *A Short Residence in Sweden, Norway, and Denmark* (Penguin, 1987); Virginia Woolf, *The Diary of Virginia Woolf*, ed. Anne Olivier Bell, vol. 3, *1925–1930* (Hogarth Press, 1980);

Virginia Woolf, *Selected Essays* (Oxford University Press, 2009); Virginia Woolf, *The Waves* (Vintage, 2004).

8. THE LITTLE GOD

W. H. Auden, *Selected Poems* (Faber, 1979); Martin Blaser, *Missing Microbes* (One World, 2014); Christian Bök, 'The Xenotext Works', 2 April 2011, https://www.poetryfoundation.org/harriet/2011/04/the -xenotext-works; Douglas Ian Campbell and Patrick Michael Whittle, *Resurrecting Extinct Species* (Palgrave, 2017); P. J. Capelotti, 'Mobile Artefacts in the Solar System and Beyond', in *Archaeology and Heritage of the Human Movement into Space*, ed. Beth Laura O'Leary and P. J. Capelotti (Springer, 2015); Denise Chow, 'On the Moon, Flags and Foot-prints of Apollo Astronauts Won't Last Forever', Space.com, 6 September 2011, https://www.space.com/12846-apollo-moon-landing-sites-flags -footprints.html; Gary Cook et al., *Clicking Clean 2017* (Greenpeace, 2016); Sarah Craske, http://www.sarahcraske.co.uk; Jason Daley, 'In a First, Archival-Quality Performances Are Preserved in DNA', *Smithsonian*, 2 October 2017, https://www.smithsonianmag.com/smart-news /two-rare-music-performances-archived-dna-180965088/; Anna Davison, 'The Most Extreme Life-Forms in the Universe', *New Scientist*, 26 June 2008; Deep Carbon Observatory, 'Life in Deep Earth Totals 15 to 23 Billion Tonnes of Carbon – Hundreds of Times More Than Humans', 10 December 2018, https://deepcarbon.net/life-deep-earth-totals-15-23 -billion-tonnes-carbon; Philip K. Dick, *The Preserving Machine and Other Stories* (Pan, 1972); James J. Elser, 'A World Awash with Nitrogen', *Science* 334, no. 6062 (2011); Andy Extance, 'How DNA Could Store All the World's Data', *Nature* 537, no. 7618 (2016); J. R. Ford et al., 'An Assess-ment of Lithostratigraphy for Anthropogenic Deposits', in *A Stratigraphic Basis for the Anthropocene*, ed. Colin Waters et al. (Geological Society of London, 2014); Michael Gillings, 'Evolutionary Consequences of Anti-biotic Use for the Resistome, Mobilome, and Microbial Pangenome', *Frontiers in Microbiology* 4, no. 4 (2013); Michael Gillings, 'Lateral Gene Transfer, Bacterial Genome Evolution, and the Anthropocene', *Annals of the New York Academy of Sciences* 1389, no. 1 (2017); Michael Gillings and Ian Paulson, 'Microbiology of the Anthropocene', *Anthropocene* 5 (2014); Michael Gillings and H. W. Stokes, 'Are Humans Increasing Bacterial Evolvability?', *Trends in Ecology and Evolution* 27, no. 6 (2012); Michael

Gillings et al., 'Ecology and Evolution of the Human Microbiota', *Genes* 6, no. 3 (2015); Michael Gillings et al., 'Using the Class 1 Integron-Integrase Gene as a Proxy for Anthropogenic Pollution', *ISME Journal* 9, no. 6 (2015); Alice Gorman, 'The Anthropocene in the Solar System', *Journal of Contemporary Archaeology* 1, no. 1 (2014); Alice Gorman, 'Culture on the Moon: Bodies in Time and Space', *Archaeologies* 12, no. 1 (2016); Clive Hamilton, 'The Theodicy of the "Good Anthropocene",' *Environmental Humanities* 7, no. 1 (2016); Robert M. Hazen et al., 'On the Mineralogy of the Anthropocene Epoch' *American Mineralogist* 102 (2017); Douglas Heaven, 'Video Stored in Live Bacterial Genome Using CRISPR Gene Editing', *New Scientist*, 12 July 2017; Myra J. Hird, 'Coevolution, Symbiosis and Sociology', *Ecological Economics* 69, no. 4 (2010); Myra J. Hird, *The Origins of Sociable Life* (Palgrave, 2009); Heinrich Holland, 'The Oxygenation of the Atmosphere and Oceans', *Philosophical Transactions of the Royal Society B* 361, no. 1470 (2006); Rowan Hooper, 'Tough Bug Reveals Key to Radiation Resistance', *New Scientist*, 25 March 2007; Gerda Horneck et al., 'Space Microbiology', *Microbiology and Molecular Biology Reviews* 74, no. 1 (2010); A. A. Imshenetsky et al., 'Upper Boundary of the Biosphere', *Applied and Environmental Microbiology* 35, no. 1 (1978); Carole Lartigue et al., 'Genome Transplantation in Bacteria: Changing One Species to Another', *Science* 317, no. 5838 (2007); Jeff Long, 'Scientists Rouse Bacterium from 250-Million-Year Slumber', *Chicago Tribune*, 19 October 2000; C. Magnabosco et al., 'The Biomass and Biodiversity of the Continental Subsurface', *Nature Geoscience* 11 (2018); Lynn Margulis, *Symbiotic Planet* (Basic Books, 1998); Lynn Margulis and Dorian Sagan, *Microcosmos* (University of California Press, 1997); Mary J. Marples, 'Life on the Human Skin', *Scientific American*, 1 January 1969; Beth Laura O'Leary, '"To Boldly Go Where No Man [*sic*] Has Gone Before": Approaches in Space Archaeology and Heritage', in *Archaeology and Heritage of the Human Movement into Space*, ed. Beth Laura O'Leary and P. J. Capelotti (Springer, 2015); Ovid, *Metamorphoses*, trans. Mary Innes (Penguin, 1955); Elizabeth Pennisi, 'Synthetic Genome Brings New Life to Bacterium', *Science* 328, no. 5981 (2010); Joseph N. Pleton, *Space Debris and Other Threats from Outer Space* (Springer, 2013); Oliver Plümper et al., 'Subduction Zone Forearc Serpentinites as Incubators for Deep Microbial Life', *PNAS* 114 (2017); David Reinsel et al., *Data Age 2025* (International Data Corporation, 2018); Ben C. Scheel et al., 'Amphibian Fungal Panzootic Causes Catastrophic and Ongoing Loss of Biodiversity', *Science* 363, no. 6434 (2019); Vaclav Smil, *The Earth's Biosphere* (MIT Press, 2002); Laura Snyder, *Eye of the Beholder* (Head of Zeus, 2015); '250 Million Year Old Bacterial Spore Comes Back to Life', Bioprocess Online, 20 October 2000, https://www.bioprocessonline .com/doc/250-million-year-old-bacterial-spore-comes-ba-0001; Nikea Ulrich et al., 'Experimental Studies Addressing the Longevity of *Bacillus subtilis* Spores – the First Data from a 500-Year Experiment', *PLoS One* 13, no. 12 (2018); Peter C. Van Wyck, *Signs of Danger: Waste, Trauma and Nuclear Threat* (University of Minnesota Press, 2005); Milton Wainwright, *Miracle Cure* (Basil Blackwell, 1990); William Whitman et al., 'Prokaryotes: The Unseen Majority', *PNAS* 95, no. 12 (1998); Pak Chung Wong et al., 'Organic Data Memory Using the DNA Approach', *Communications of the ACM* 46, no. 1 (2003); Shoshuke Yoshida et al. 'A Bacterium That Degrades and Assimilates Poly(ethylene terephthalate)', *Science* 351, no. 6278 (2016); Jan Zalasiewicz, 'The Extraordinary Strata of the Anthropocene', *Environmental Humanities*, ed. Serpil Oppermann and Serenella Iorvino (Rowman and Littlefield, 2017); Jan Zalasiewicz et al., 'The Mineral Signature of the Anthropocene in Its Deep-Time Context', in *A Stratigraphic Basis for the Anthropocene*, ed. Colin Waters et al. (Geological Society of London, 2014); Jan Zalasiewicz et al., 'The Technofossil Record of Humans', *Anthropocene Review* 1, no. 1 (2014); Eric Zettler et al., 'Life in the "Plastisphere": Microbial Communities on Plastic Marine Debris', *Environmental Science and Technology* 47, no. 13 (2013); Young-Guan Zhu et al., 'Microbial Mass Movements', *Science* 357, no. 6356 (2017); Sergey A. Zimov, 'Pleistocene Park: Return of the Mammoth's Ecosystem', *Science* 308, no. 5723 (2005); Sergey A. Zimov et al., 'Permafrost and the Global Carbon Budget', *Science* 312 (2006).

CODA: SEEING THE NEW WORLD

Italo Calvino, *Collection of Sand*, trans. Martin McLaughlin (Penguin, 2013); Patricia L. Corcoran et al., 'An Anthropogenic Marker Horizon in the Future Rock Record', *GSA Today* 24, no. 6 (2014); Paul Valéry, *Eupalinos, or The Architect*, trans. William McCausland Stewart (Oxford University Press, 1932).

ACKNOWLEDGMENTS

This book bears traces of many acts of kindness.

Wherever I have travelled in search of future fossils, I have met the most welcoming guides and hosts: Alison Sheridan; Elle Leane, Andrew Moy and Meredith Nation; Jody Webster, Madhavi Patterson and Belinda Dechnik; Pasi Tuoma and Anne Kontula; Christina Fredengren and Lena Kautsky; Christine Hansen, Kerstin Johannesson and Matthias Obst; Michael Gillings and Sasha Tetu; and Christian Bök. Without their generosity, I'd have been left with a very thin book indeed. Vincent Ialenti talked me through life 450 metres underground, and Jan Zalasiewicz helped me to imagine one-million-year-old plastic.

I began thinking about how to tell the story of future fossils during a three-month stay in Australia with my family, funded by the Leverhulme Trust. I acknowledge the precedence of the traditional owners whose country I visited in researching *Footprints*, and pay respect to their continuing connection to the land and waters of which they are custodians: the Gadigal of the Eora Nation, the Kuku Yalanji and the Mirrar Gudjeihmi. I'm grateful to Thom van Dooren for helping arrange my fellowship at the University of New South Wales, and to Iain McCalman, Astrida Neimanis, and their respective families for welcoming us so warmly in Sydney. Thanks, too, to Julian Barry for letting us tour the Northern Territory in his car, and to Janet Black, Vicki Kincaid and Laura Tomlinson at the University of Edinburgh for logistical support during all my travels.

I have been very lucky to have so many people offer encouragement, often just when I needed it most. Thank you to Esa Aldegheri, Rebecca Altman, James Bradley, Simon Cooke, Tim Dee, Peter Dorward, my parents Lynn and Ian, Tom Killingbeck, Robert Macfarlane, Max Porter and Kate Rigby for your kind words and enthusiasm. Two very fine people I am lucky to call my friends, Gavin Francis and Ben White, offered insightful observations about the first draft that helped me see my way to something much better. Special thanks to my friends in the 'deep time' reading group at the University of Edinburgh: Michelle Bastian, Emily Brady, Franklin Ginn, Jeremy Kidwell and Andrew Patrizio. Many of the stories I've told here were seeded in our conversations.

Receiving the 2017 Giles St Aubyn Award from the Royal Society of Literature was a tremendous boost to my confidence, and I am immensely grateful to all at the RSL and to the judges for their support.

I owe a debt to all those who gave me an opportunity to tell a story I feel so passionate about: my thanks to Sally Davies at *Aeon*, who first invited me to write about the deep future, and Lettice Franklin, for the faith she showed in commissioning *Footprints*. I have been extremely fortunate to work with Zoë Pagnamenta, with Nicholas Pearson at Fourth Estate and with Eric Chinski and

ACKNOWLEDGMENTS

Julia Ringo at Farrar, Straus and Giroux. I couldn't have imagined a finer, more committed team of editors, and I have benefited enormously from their sensitivity and belief in the strange tales I presented them with.

My agent, the wonderful Carrie Plitt, has been a guide and friend from start to finish, and made me see that perhaps I had a story to tell after all. Thank you, Carrie.

To Rachel, whom I admire more than anyone I have ever met, I owe more than I can say.

To Isaac and Annie, this is for your future.